# oxford **maths**

## for australian schools

# contents

# OXFORD

## UNIVERSITY PRESS

## Practice

**1** What is the value of the red digit in each number?

a    734 815

b    62 759 377

c    6 219 730

d    13 585 104

e    5 487 800

f    218 819 999

**2** Write the numbers from question 1 in words.

a    _____

b    _____

c    _____

d    _____

e    _____

f    _____

**3** Expand these numbers. The first one has been started for you.

a    413 629:

     400 000 + _____

b    3 746 123:

     _____

c    52 065 350:

     _____

d    43 200 806:

     _____

e    680 405 020:

     _____

Oxford University Press

# Challenge

**1** Look at the digit cards.

| 5 | 2 | 8 | 7 | 9 | 3 | 1 | 0 |

**a** What is the **smallest** whole number that can be made using all the cards if the 0 is in the millions place?

_____

**b** What is the **largest** number that can be made if the digit 1 is in the tens of millions place?

_____

**c** What is the **smallest** number that can be made if the 9 is the last digit and is used as nine-tenths?

_____

**d** What is the **largest** number that can be made if the digits 7, 8 and 9 are used as decimals?

_____

**2** Match each number to the correct spike abacus.

**A**

**B**

**C**

**D**

**a** 31 084 509 matches to spike abacus _____

**b** Three hundred and ten million, eighty thousand, four hundred and fifty-nine matches to spike abacus _____

**c** 310 084 509 matches to spike abacus _____

**d** Three hundred and one million, eight hundred and four thousand, five hundred and nine matches to spike abacus _____

**3** Write the answers to the following using digits in full.

**a** Increase $375K by $20 000 _____

**b** Decrease $1.5M by $200 000 _____

**c** Decrease $950K by $40 000 _____

**d** Increase $1.25M by $250 000 _____

1. In 2016 the number of cats kept as pets in Australia, rounded to the nearest 100 000, was 3.9M.

   Some of the following could be the exact number. Circle the possible answers.

   3 487 245     3 878 224     4 900 010     3 995 110

   3 949 999     3 898 100     3 918 885     4 000 001

2. In a survey carried out a few years ago, the dog population of China was around 22M. The red arrow shows the actual number of dogs.

   5 000 000

   Circle any of the following numbers that could more accurately represent the actual dog population of China.

                                                    19.919M

                         22.909M                                              20.899M

   23.009M

              21.142M                    24.952M                  22.985M

3. The population of Brisbane recently reached 2.3M, rounded to the nearest hundred thousand. The exact number can be found by using each of the following digits once:

   1     2     3     5     6     7     9

   Make a list of numbers that could represent the actual population of Brisbane. (There are more than 20 possible correct answers.)

Oxford University Press

## Practice

**1** Lightly shade all the square numbers on this hundred grid.

| 1 | 2 | 3 | 4 | 5 | 6 | 7 | 8 | 9 | 10 |
|---|---|---|---|---|---|---|---|---|----|
| 11 | 12 | 13 | 14 | 15 | 16 | 17 | 18 | 19 | 20 |
| 21 | 22 | 23 | 24 | 25 | 26 | 27 | 28 | 29 | 30 |
| 31 | 32 | 33 | 34 | 35 | 36 | 37 | 38 | 39 | 40 |
| 41 | 42 | 43 | 44 | 45 | 46 | 47 | 48 | 49 | 50 |
| 51 | 52 | 53 | 54 | 55 | 56 | 57 | 58 | 59 | 60 |
| 61 | 62 | 63 | 64 | 65 | 66 | 67 | 68 | 69 | 70 |
| 71 | 72 | 73 | 74 | 75 | 76 | 77 | 78 | 79 | 80 |
| 81 | 82 | 83 | 84 | 85 | 86 | 87 | 88 | 89 | 90 |
| 91 | 92 | 93 | 94 | 95 | 96 | 97 | 98 | 99 | 100 |

**2** Underline the sentence that is a statement of fact:

All square numbers are odd.

All square numbers are even.

There is an equal number of odd and even square numbers.

There are more odd square numbers than even square numbers.

There are more even square numbers than odd square numbers.

**3** Circle any of the options below that complete the following sentence correctly.

If you add together the first five odd numbers, the answer is the same as:

$5 \times 5$ $\qquad$ $1 \times 5$ $\qquad$ $5^2$ $\qquad$ $5 \times 2$ $\qquad$ $2 \times 5$

**4** Circle the first 10 triangular numbers on the hundred grid in question 1.

**5** Which numbers on the hundred grid are both triangular and square numbers?

_____

**1** The 11th square number is 121. The sum of which consecutive 11 odd numbers is 121?

_____

**2** **a** What is $13^2$? _____

**b** Write the addition sequence that will give the same answer as $13^2$.

_____

**3** The 10th triangular number is 55. The 11th triangular number can be found by adding what to 55?

_____

**4** This pattern of beads shows the third triangular number, which is 6.

**a** Draw a pattern of beads showing the 11th triangular number.

$1 + 2 + 3 = 6$

**b** Write the addition fact below the pattern using numbers that are added together to show the 11th triangular number.

**1** Drawing and counting the dots in a small triangular number, such as 6, is simple. Larger triangular numbers can take a long time to work out in this way. However, there is a mathematical way of calculating **any** triangular number.

To understand this, we'll look at the second triangular number (2nd TN).

2nd TN

- Since it is the 2nd TN, we begin by drawing the dots as a right-angled triangle that is 2 dots high and 2 dots wide.

- Next, we **double the number of dots and make a rectangle**.

- The rectangle is one dot wider than it is high (2 dots high × (2 + 1) dots wide). We can write this more simply as:

*The rectangle is 2 × (2 + 1) = 2 × 3 = 6 dots.*

- We **doubled** the number of dots to make the rectangle, so we **halve** the number of dots to find the triangular number.

- The 2nd TN is 2 × (2 + 1) ÷ 2 = 2 × 3 ÷ 2 = 3 dots.

This will work for **any** triangular number. We used 2 for the 2nd TN, so we use 3 for the 3rd TN.

a   Look at the rectangle and complete the calculation:

3rd TN = 3 × (3 + 1) ÷ 2

= _____

3rd TN

3

×        3 + 1

b   Use the diagrams and the calculation to prove that the fifth triangular number is 15.

_____

_____

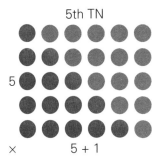

5th TN

5

×        5 + 1

**2** Use the rule for calculating triangular numbers to work out as many triangular numbers as you wish on a separate piece of paper. You could begin by calculating the value of the 10th and 20th triangular numbers.

### Practice

**1** Complete the chart to show the factors of the prime and composite numbers between 21 and 30.

Remember: A *prime* number has just two factors—1 and itself. All other numbers are *composite* numbers

| Numbers | Factors (numbers it can be divided by) | How many factors? | Is it prime or composite? |
|---|---|---|---|
| 21 | | | |
| 22 | | | |
| 23 | | | |
| 24 | | | |
| 25 | | | |
| 26 | | | |
| 27 | | | |
| 28 | | | |
| 29 | | | |
| 30 | | | |

**2** List the factors of 12. _____

**3** a Which number between 21 and 30 has the same amount of factors as 12? _____

   b Which numbers between 21 and 30 have more factors than 12? _____

**4** There is only one even prime number. What is it? _____

**5** a How many factors does the square number on the chart above have? _____

   b Name a square number lower than 10 that has the same number of factors as the square number on the chart above. _____

Oxford University Press

**1** Fill in the gaps to show the prime factors of the following:

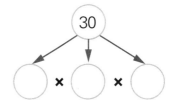

**a** The prime factors of 14 are:

_____

**b** The prime factors of 77 are:

_____

**c** The prime factors of 30 are:

_____

Remember: 2 × 2 can be written as $2^2$ and 2 × 2 × 2 can be written as $2^3$.

**2** Fill in the gaps.

**a**

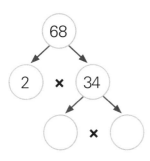

68 = 2 × 2 × _____

68 = $2^{\square}$ × _____

**b**

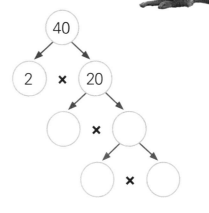

40 = _____ × _____ × _____ × _____

40 = $2^{\square}$ × _____

**c**

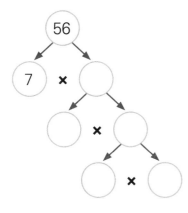

56 = _____ × _____ × _____ × _____

56 = _____ × $2^{\square}$ ___

**d**

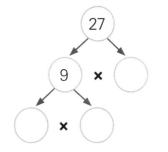

27 = _____ × _____ × _____

27 = $_{\square}$ _____

# Mastery

**1** How do you know that 37 is a prime number?

_____

**2** Complete the gaps in the table to show each number as a product of its prime factors. Also use the index form.

| | Prime factors | Index form | Number |
|---|---|---|---|
| a | $2 \times 2 \times 2 \times 3$ | $2^3 \times 3$ | |
| b | | $2^3 \times 5^2$ | 200 |
| c | $2 \times 2 \times 2 \times 2 \times 5$ | | |
| d | | | 84 |

**3** a _____$^2$ = 16    b _____$^3$ = 8    c _____$^2$ = 49    d _____$^3$ = 27

**4** Another way of writing question 3a would be, "Find the square root of 16" or "$\sqrt{16}$".

a $\sqrt{36}$ = _____    b $\sqrt{64}$ = _____    c $\sqrt{81}$ = _____

**5** We can find the square root of certain, less-common numbers if the number is written as a **product of its prime factors**. Look at the first example and complete the second one.

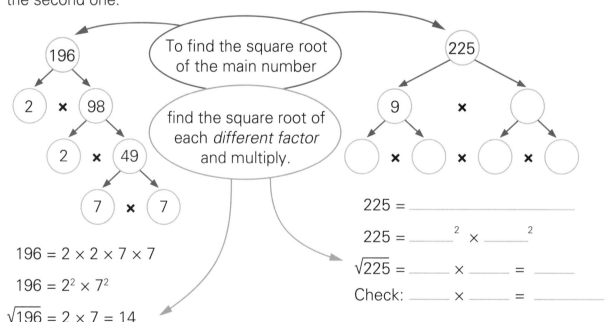

To find the square root of the main number

find the square root of each *different factor* and multiply.

$196 = 2 \times 2 \times 7 \times 7$

$196 = 2^2 \times 7^2$

$\sqrt{196} = 2 \times 7 = 14$

Check: 14 x 14 = 196

$225 = $ _____

$225 = $ _____$^2$ × _____$^2$

$\sqrt{225} = $ _____ × _____ = _____

Check: _____ × _____ = _____

Oxford University Press

### Practice

**1** You could use the split strategy to solve these:

a   231 + 427 _____

b   216 + 443 _____

c   536 + 822 _____

d   1823 + 636 _____

e   4285 + 714 _____

f   4724 + 8265 _____

**2** You could use the rounding (compensation) strategy to solve these:

*Some of these involve trading.*

a   638 − 399 _____

b   427 + 302 _____

c   362 + 698 _____

d   768 − 501 _____

e   857 − 290 _____

f   998 + 403 _____

**3** Explain the method that you use to solve the following:

a   551 + 348 = _____

b   1472 − 299 = _____

c   4350 + 4350 = _____

d   24 746 − 520 = _____

# Challenge

**1** Here is a list of five of the most expensive paintings ever sold. Decide how to round the numbers. Explain how you have rounded them.

|   | Painting title | Price | Artist | Rounded price | Rounded to the nearest |
|---|---|---|---|---|---|
| a | *Salvator Mundi* (painted around 1500) | $450 300 000 | Leonardo da Vinci | | |
| b | *Les Femmes d'Alger* (painted in the 1950s) | $179 000 000 | Pablo Picasso | | |
| c | *Nu Couché* (painted in 1917/18) | $170 400 000 | Amedeo Modigliani | | |
| d | *Three Studies of Lucian Freud* (painted in the 1960s) | $142 400 000 | Francis Bacon | | |
| e | *The Scream* (painted in the 1890s) | $119 922 500 | Edvard Munch | | |

**2** **a** If the same person bought paintings **a** and **d**, would the amount be closer to $592M or $593M? _____

**b** How much less was painting **c** than painting **b**? _____

**c** How much less than $120M was painting **e**? _____

**d** What was the total paid for paintings **c** and **d**? _____

**3** Estimate the answer to each problem below, then check on a calculator. Make sure your answer is close to the calculator answer. If it isn't, check what went wrong.

|   | Problem | Round the numbers | Estimate the answer | Calculator answer |
|---|---|---|---|---|
| e.g. | 4195 + 3030 | 4000 + 3000 | 7000 | 7225 |
| a | 3085 + 1877 | | | |
| b | 6018 – 3119 | | | |
| c | 35 056 + 4023 | | | |
| d | 29 856 – 20 028 | | | |
| e | 400 247 + 49 588 | | | |
| f | 398 108 – 197 434 | | | |
| g | 5 498 858 – 2 889 917 | | | |

Oxford University Press

1   Two adults and two children go on a holiday. The flights cost $199 for each child and $299 for each adult. The accommodation costs $1299 and they spend $310 on food.

   a   Is the price of the holiday closest to $2550, $2600, $2650 or $2700?

   _____

   b   What is the exact price of the holiday? _____

2   You may need to use a calculator to solve this problem.

   a   Audrey wrote the nine digits for her friend like this:

   # 123456789

   She asked, "What would you need to add to the number to get this answer?":

   # 987654321

   b   Write the number that would need to be added. Write your answer in digits, with the correct spacing, and in words.

   _____

3   A theatre holds around 500 people. At the end of eight shows, exactly 4000 people have attended, but no two performances had exactly the same number in the audience. Show in the table how many people could have been at each performance.

| Performance | Number of people |
|---|---|
| 1 | |
| 2 | |
| 3 | |
| 4 | |
| 5 | |
| 6 | |
| 7 | |
| 8 | |
| **Total** | |

## Practice

**1** Set out these addition problems vertically. Look for patterns in the answers.

**a**  107 + 16

+ _____
_____

**b**  387 + 69

+ _____
_____

**c**  425 + 27 + 337

+ _____
_____

**d**  784 + 354 + 96

+ _____
_____

**e**  1786 + 10 528 + 31

+ _____
_____

**f**  89 316 + 5892 + 28 248

+ _____
_____

**g**  758 306 + 392 048 + 84 213

+ _____
_____

**h**  8 235 946 + 428 637 + 1 665 736 + 2 015 359

+ _____
_____

**i**  845 386 + 57 214 936 + 35 678 523 + 22 581 099 + 7 136 845

+ _____
_____

Remember: Keep the digits in the correct columns.

Oxford University Press

# Challenge

Did you know that in Australia there are almost half a million kilometres of unpaved roads?

This list shows the total length of paved and unpaved roads in 9 countries. The list is in order from the country with the greatest length of paved roads, to the country with the smallest length of paved roads.

| | Country | Length of paved roads (km) | Length of unpaved roads (km) | Total length of roads (km) |
|---|---|---|---|---|
| 1 | USA | 4 165 110 | 2 265 256 | |
| 2 | India | 1 603 705 | 1 779 639 | |
| 3 | China | 1 515 797 | 354 864 | |
| 4 | Japan | 925 000 | 258 000 | |
| 5 | Russia | 738 000 | 133 000 | |
| 6 | Spain | 659 629 | 6 663 | |
| 7 | Canada | 415 600 | 626 700 | |
| 8 | Australia | 336 962 | 473 679 | |
| 9 | Brazil | 96 353 | 1 655 515 | |

**1**

**a** Without using a calculator, find the total length of roads for each country and then fill in the last column of the table.

**b** The order changes when the total length of roads has been calculated. Re-order the list, from the country with the greatest total length of all roads, to the country with the smallest total length of all roads.

**1** The answer is 123321. There are two 5-digit addends. Write three different sums that would give the correct answer.

**2** Find four numbers that can be added together to give an answer of 123 456 789. None of the numbers should have less than four digits.

**3** The answer is 12 046. Use the digits 1, 0, 5 and 7 to write the sum. You can use each digit more than once. You may wish to use a calculator.

**4** The answer is 987 654 321. There are four addends: one has nine digits; one has eight digits; one has seven digits and one has six digits. Write two different sums that would give the correct answer.

Oxford University Press

## Practice

**1** Look for a connection between each subtraction and the one before it.

a
| 3 | 6 | 9 | 7 | 4 | 8 |
|---|---|---|---|---|---|
| − | 2 | 4 | 8 | 5 | 3 | 6 |

b
| 3 | 1 | 8 | 2 | 5 | 9 |
|---|---|---|---|---|---|
| − | | 8 | 5 | 9 | 3 | 6 |

c
| 4 | 7 | 2 | 7 | 9 | 1 |
|---|---|---|---|---|---|
| − | 1 | 2 | 9 | 3 | 5 | 7 |

d
| 7 | 2 | 2 | 8 | 6 | 1 |
|---|---|---|---|---|---|
| − | 2 | 6 | 8 | 3 | 1 | 6 |

e
| 9 | 7 | 1 | 4 | 8 | 5 |
|---|---|---|---|---|---|
| − | 4 | 0 | 5 | 8 | 2 | 9 |

f
| 7 | 9 | 2 | 3 | 9 | 8 |
|---|---|---|---|---|---|
| − | 1 | 1 | 5 | 6 | 3 | 1 |

g
| 1 | 1 | 4 | 6 | 1 | 2 | 3 |
|---|---|---|---|---|---|---|
| − | | 3 | 5 | 8 | 2 | 4 | 5 |

h
| 1 | 3 | 8 | 1 | 3 | 5 | 6 |
|---|---|---|---|---|---|---|
| − | | 4 | 8 | 2 | 3 | 6 | 7 |

i
| 1 | 1 | 2 | 4 | 8 | 5 | 4 |
|---|---|---|---|---|---|---|
| − | | 1 | 3 | 4 | 9 | 5 | 6 |

j
| 1 | 1 | 2 | 8 | 1 | 6 | 8 |
|---|---|---|---|---|---|---|
| − | | 2 | 4 | 9 | 3 | 8 | 1 |

**2** Write an algorithm to show that 343 965 subtracted from another number gives an answer of 767 676.

**1** These subtractions involve zeros in the top line. Look for a connection between each subtraction and the one before it.

a

| | 4 | 0 | 4 | 6 | 3 | 1 |
|---|---|---|---|---|---|---|
| − | | 8 | 3 | 3 | 1 | 0 |

b

| | 7 | 2 | 0 | 3 | 0 | 5 |
|---|---|---|---|---|---|---|
| − | 2 | 8 | 7 | 8 | 7 | 3 |

c

| | 6 | 3 | 0 | 0 | 4 | 2 | 9 | 6 |
|---|---|---|---|---|---|---|---|---|
| − | 1 | 9 | 7 | 8 | 9 | 9 | 7 | 5 |

d

| | 8 | 2 | 0 | 5 | 0 | 6 | 0 | 7 |
|---|---|---|---|---|---|---|---|---|
| − | 2 | 7 | 7 | 2 | 5 | 1 | 7 | 5 |

e

| | 9 | 4 | 0 | 0 | 1 | 0 | 0 | 5 |
|---|---|---|---|---|---|---|---|---|
| − | 2 | 8 | 5 | 6 | 4 | 4 | 6 | 2 |

f

| | 8 | 0 | 0 | 0 | 3 | 0 | 0 | 0 |
|---|---|---|---|---|---|---|---|---|
| − | | 3 | 4 | 5 | 5 | 3 | 4 | 6 |

g

| | 9 | 0 | 1 | 4 | 5 | 0 | 9 | 4 |
|---|---|---|---|---|---|---|---|---|
| − | | 2 | 4 | 8 | 6 | 3 | 2 | 9 |

h

| | 1 | 0 | 0 | 2 | 0 | 0 | 0 | 8 | 0 |
|---|---|---|---|---|---|---|---|---|---|
| − | | | 1 | 4 | 3 | 0 | 2 | 0 | 4 |

i

| | 5 | 0 | 0 | 5 | 0 | 0 | 5 | 0 | 0 |
|---|---|---|---|---|---|---|---|---|---|
| − | 3 | 7 | 7 | 0 | 4 | 6 | 1 | 7 | 9 |

j

| | 8 | 0 | 8 | 0 | 8 | 0 | 8 | 0 | 8 |
|---|---|---|---|---|---|---|---|---|---|
| − | 2 | 6 | 4 | 8 | 6 | 8 | 4 | 6 | 3 |

**2** Write a 9-digit subtraction for which the answer is 123 456 789. The number on the top line of the subtraction should have nine digits and include six zeros.

Oxford University Press

**1** Avocados are grown in Australia. In 2019, 79 533 tonnes of them were grown.
By how much did this number increase from the 67 600 tonnes that were produced in 2016? _____

**2** Over 79 000 tonnes of Australian avocados seems like a lot, but it is not much compared to the top five avocado-producing countries.

This table compares the number of avocados each of the top five countries produced in 2019 and in 2016.

| Country | Avocado production in tonnes (2019) | Avocado production in tonnes (2016) |
|---|---|---|
| Mexico | 2 300 889 t | 1 889 354 t |
| Dominican Republic | 661 626 t | 601 349 t |
| Peru | 535 911 t | 455 394 t |
| Colombia | 535 021 t | 294 389 t |
| Indonesia | 461 613 t | 304 938 t |

a   Which country in the table produced the least number of avocados in 2016?

_____

_____

b   By how many tonnes did the Dominican Republic's avocado production increase between 2016 and 2019? _____

_____

c   What was the difference between the numbers produced in Indonesia in 2016 and 2019? _____

_____

d   What was the difference between the numbers produced in the top country and in Australia in 2016? _____

_____

e   By how many tonnes was the number of avocados produced in the top country greater than the total number produced by the other four countries in 2019?

_____

_____

f   By how much was the total produced by the five countries greater in 2019 than it was in 2016? _____

_____

**1** Fill in the table to multiply by 10, by 100 and by 1000.

|   |   | × 10 | × 100 | × 1000 |
|---|---|---|---|---|
| a | 43 | | | |
| b | 87 | | | |
| c | 235 | | | |
| d | $7.75 | | | |
| e | 250 | | | |
| f | 36 | | | |
| g | $0.25 | | | |
| h | $10.15 | | | |

Remember: When you multiply by 10 the digits move one place bigger. When you divide by 10 the digits move one place smaller.

**2** Fill in the table to divide by 10 and then write the multiplication-fact partner. There are two reminders for you about using mental strategies to divide by 10.

|   |   | ÷ 10 | Write the multiplication-fact partner |
|---|---|---|---|
| e.g. | 230 | 23 | 23 × 10 = 230 |
| e.g. | 25 | 2.5 | 2.5 × 10 = 25 |
| a | 350 | | |
| b | 3600 | | |
| c | 5000 | | |
| d | $12.50 | | |
| e | 73 | | |

**3** Fill in the table to divide by 100 and then write the multiplication-fact partner. There are two reminders for you about using mental strategies to divide by 100.

|   |   | ÷ 100 | Write the multiplication-fact partner |
|---|---|---|---|
| e.g. | 900 | 9 | 9 × 100 = 900 |
| e.g. | $325 | $3.25 | $3.25 × 100 = $325 |
| a | 800 | | |
| b | $255 | | |
| c | 7000 | | |
| d | 15 000 | | |
| e | 7250 | | |

Oxford University Press

**1** Use the doubles strategy to multiply by multiples of 10.

|       |     | × 10 | × 20 | × 40 | × 80 |
|-------|-----|------|------|------|------|
| e.g.  | 24  | 240  | 480  | 960  | 1920 |
| a     | 11  |      |      |      |      |
| b     | 14  |      |      |      |      |
| c     | 21  |      |      |      |      |
| d     | 35  |      |      |      |      |
| e     | 60  |      |      |      |      |

**2** Use halving strategies to divide by multiples of 10.

|       |        | ÷ 10 | ÷ 20 | ÷ 40 | ÷ 80 |
|-------|--------|------|------|------|------|
| e.g.  | 1200   | 120  | 60   | 30   | 15   |
| a     | 800    |      |      |      |      |
| b     | 4000   |      |      |      |      |
| c     | 720    |      |      |      |      |
| d     | 12 000 |      |      |      |      |
| e     | 4800   |      |      |      |      |

**3** Use a time-saving mental strategy to solve the following. Be ready to explain how you got the answers.

*The short cut that works for **you** is the best short cut!*

a   $25 \times 20$ _____

b   $24 \times 40$ _____

c   $15 \times 50$ _____

d   $14 \times 40$ _____

e   $25 \times 30$ _____

f   $125 \times 20$ _____

g   $25 \times 60$ _____

h   $42 \times 40$ _____

i   $19 \times 20$ _____

j   $\$1.30 \times 50$ _____

k   $\$2.75 \times 20$ _____

l   $448 \div 4$ _____

m   $450 \div 5$ _____

n   $1800 \div 5$ _____

**4** Explain a mental strategy that you would use to multiply $22 \times 14$.

_____

_____

**1** If you know that $9 \times 7 = 63$, you also know that $90 \times 70 =$ (Circle one):

630                    6030                    6300                    63 000

**2** **a** $600 \times 40 = 24\ 000$. What is the connection between the number of zeros in the two numbers being multiplied and in the answer?

_____

**b** Test whether the connection you noticed works for other numbers.

**3** **a** Circle the sentences that would give the correct answer to $648 \div 20$:

- You could halve 648 and move the digits one place smaller.
- You could divide 648 by 10 and then halve the answer.
- You could halve 648 and move the digits one place bigger.
- You could divide 648 by 2 and then by 10.

**b** Choose a strategy to divide 648 by 20. Explain why you chose that strategy.

**4** A school needs 30 new outdoor seats. The first supplier sells five seats for $107.50. A second offers a crate of 32 seats for $624. If you were the principal, which deal would you choose? Give a reason for your answer.

Oxford University Press

## Practice

**1** Use short or extended multiplication to solve the following:

a
```
    4  2  6
×         2
_____
```

b
```
    3  1  9
×         3
_____
```

c
```
    1  3  6
×         5
_____
```

d
```
    2  4  6
×         3
_____
```

e
```
  1  7  3  9
×          4
_____
```

f
```
  1  2  8  5
×          6
_____
```

g
```
  2  3  4  5
×          4
_____
```

h
```
  1  6  3  7
×          7
_____
```

i
```
  2  3  2  8
×          6
_____
```

j
```
  2  7  4  5
×          7
_____
```

k
```
  4  0  0  5
×          8
_____
```

l
```
  6  3  0  9
×          9
_____
```

m
```
  2  0  8  0
×          7
_____
```

**2** Six people go overseas. The cost for each is $3078. How much do they pay altogether?

**3** A builder sells seven new houses. The price for the first six is $349 845 each. The last one is $12 000 cheaper. How much does the builder get for all seven houses?

# Challenge

**1** Multiplying by a multiple of 10

*Remember: Putting in the zero moves everything over one place for you.*

**a**
```
    2 6
×   2 0
───────
      0
```

**b**
```
    3 4
×   2 0
───────
```

**c**
```
    2 7
×   3 0
───────
```

**d**
```
    5 3
×   4 0
───────
```

**e**
```
    2 7
×   6 0
───────
```

**f**
```
    4 4
×   5 0
───────
```

**g**
```
  1 7 3
×   3 0
───────
```

**h**
```
  2 5 3
×   6 0
───────
```

**i**
```
  6 9 4
×   7 0
───────
```

**j**
```
1 6 2 8
×     5 0
─────────
```

**k**
```
2 5 9 5
×     6 0
─────────
```

**l**
```
2 5 8 3
×     9 0
─────────
```

**2** Multiplying by two digits

*Multiplying by two digits is like doing two multiplications in one.*

**a**
```
    2 4
×   2 5
───────
          ← 24 × 5
+       0 ← 24 × 20
───────
───────
```

**b**
```
    3 5
×   2 6
───────

+
───────
```

**c**
```
    4 8
×   1 8
───────

+
───────
```

**d**
```
  1 5 8
×   2 4
───────

+
───────
```

**e**
```
  3 0 8
×   6 5
───────

+
───────
```

**f**
```
  2 9 7
×   4 8
───────

+
───────
```

Oxford University Press

**1** Did you know that the five largest sea animals are all whales?

Use the lengths and weights of the various whales in the table to answer the questions below.

a Which type of whale has a mass that is closest to twice that of another whale?

_____

b What would the total mass of 23 northern right whales be?

_____

c If 15 blue whales were put nose to tail, what would the total length be? _____

|  | Mammal | Length | Weight |
|---|---|---|---|
| 1 | Blue whale | 33.5 m | 137 000 kg |
| 2 | Bowhead whale | 20.2 m | 86 000 kg |
| 3 | Northern right whale | 18.35 m | 77 700 kg |
| 4 | Fin whale | 25.5 m | 63 400 kg |
| 5 | Sperm whale | 18.35 m | 43 700 kg |

d Which would have a greater mass, 26 fin whales or 12 blue whales? _____

e 15 northern right whales (nose to tail) would stretch the same distance as 15 sperm whales, but how much heavier would they be? _____

**2** When some numbers are multiplied together they start to make an interesting pattern. Square numbers can make an interesting pattern, but you may need a calculator to find them. For example, 111 × 111 makes a pattern that starts with 123 … Try it for yourself, then continue the pattern. Your calculator might not be big enough, but if you get all the way to the square of 111 111 111, the answer is 12 345 678 987 654 321.

If you square 101 010 the answer is ten, twenty, thirty, twenty, ten, zero—or ten billion, two hundred and three million, twenty thousand, one hundred (10 203 020 100).

Use a calculator and spare paper to investigate patterns with multiplications. For example, you've seen that 11 × 11 = 121. Now try 111 × 11. And 1111 × 11. What other patterns can you find? Don't forget to write down the answers for others to see.

**Practice**

**1** Complete the following, writing the reminder in two ways.

*The remainder for 25 ÷ 4 can be written as 6 r1 or 6 $\frac{1}{4}$.*

a   13 ÷ 4 _____

b   31 ÷ 4 _____

c   33 ÷ 8 _____

d   64 ÷ 9 _____

e   75 ÷ 8 _____

f   59 ÷ 7 _____

g   74 ÷ 9 _____

h   59 ÷ 6 _____

**2** Complete each algorithm, showing the remainder in two ways.

a   4 )5 6 7          4 )5 6 7

b   5 )3 7 2          5 )3 7 2

c   6 )2 0 6          6 )2 0 6

d   7 )7 4 5          7 )7 4 5

e   8 )2 5 7 0        8 )2 5 7 0

f   9 )1 7 5 4        9 )1 7 5 4

g   4 )4 1 3 9        4 )4 1 3 9

h   7 )3 3 0 9        7 )3 3 0 9

**3** Re-write using the )‾ symbol. Write the remainder using the "r_" format.

a   1286 ÷ 5

b   1844 ÷ 3

c   2277 ÷ 6

d   5826 ÷ 9

e   8046 ÷ 4

f   7055 ÷ 7

Remember: Put a decimal point and a zero after the number to be divided.

**1** Show the remainder as a decimal. Use one or two decimal places.

a $4\overline{)713.00}$   b $5\overline{)818}$   c $2\overline{)907}$

d $6\overline{)747}$   e $4\overline{)2549}$   f $8\overline{)9370}$

g $6\overline{)34383}$   h $5\overline{)71008}$   i $8\overline{)27830}$

**2** Rewrite the following using the $\overline{)}$ symbol. Show the remainder as a decimal. Stop after two decimal places, rounding the final decimal place where necessary.

a   $2539 \div 4$   b   $7453 \div 3$   c   $8467 \div 6$

d   $58\,246 \div 8$   e   $91\,347 \div 9$   f   $82\,452 \div 7$

g   $90\,236 \div 8$   h   $54\,371 \div 9$   i   $25\,385 \div 7$

**3** Solve the following problem, dealing with the remainder in the most appropriate way. A machine makes 23 455 pencils in a day. Each box holds eight pencils. How many boxes of pencils are made?

**1** Finlay's scores in six cricket games are 21, 85, 4, 47, 59 and 101.

    **a** What is his mean average score (to two decimal places)? _____

    **b** In the seventh game he scores zero. What is his new mean average? _____

    **c** After the eighth match his mean average is higher than after six games. How many runs might he get in the eighth game? _____

**2** The TV weather person said that the average temperature in the last week of the school holidays had been 31°C. The table shows the actual temperatures for three of the days. Fill in the table to show what the other temperatures might have been.

| Day | Temperature |
| --- | --- |
| Sunday | |
| Monday | 31°C |
| Tuesday | |
| Wednesday | |
| Thursday | 34°C |
| Friday | 29.5°C |
| Saturday | |

**3** In Australia, people eat quite a lot of potatoes, but not as much as in some other countries.

This table shows the four countries in which the most potatoes are eaten. It also shows the amount eaten in Australia.

| Country | Amount of potatoes per person per week | Amount of potatoes per person per day |
| --- | --- | --- |
| Belarus | 3420.41 g | |
| Poland | 2616.11 g | |
| Russia | 2513.91 g | |
| Ireland | 2436.07 g | |
| Australia | 1015.24 g | |

    **a** Work out the amount of potatoes eaten per person per day in each of the countries and fill in the final column.

    **b** Research the top five potato-producing countries, and record your findings in the table. Do you think these countries will be the same as those in the table above?

| Country | Potato production per year in tonnes |
| --- | --- |
| | |
| | |
| | |
| | |
| | |

Oxford University Press

## Practice

**1** Show the following operations on the number lines. Write the number sentence for each operation. The first number line has been started for you.

**a** Increase −4 by 5.

Number sentence: _____

**b** Decrease 3 by 5.

Number sentence: _____

**c** Decrease 6 by 8.

Number sentence: _____

**d** Decrease 0 by 7.

Number sentence: _____

**e** Increase −6 by 9.

Number sentence: _____

**f** Increase −2 by 7.

Number sentence: _____

**g** Decrease 1 by 7.

Number sentence: _____

**h** Increase 2 by 5.

Number sentence: _____

**i** Increase 0 by 6.

Number sentence: _____

**j** Decrease 7 by 11.

Number sentence: _____

**k** Increase −7 by 11.

Number sentence: _____

**l** Increase 0 by 8.

Number sentence: _____

---

**2** What would a calculator show if you pressed the following keys?

**a** $5 - 10 =$ _____

**b** $-10 + 11 =$ _____

**c** $25 - 50 =$ _____

**d** $30 - 100 =$ _____

**e** $-75 + 100 =$ _____

**f** $57 - 60 =$ _____

**1** Find the counting number for the number lines, then fill in the missing numbers.

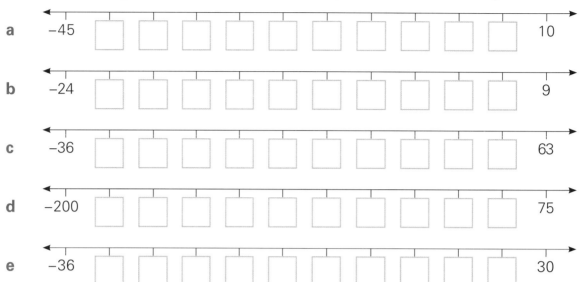

a $-45$ ☐ ☐ ☐ ☐ ☐ ☐ ☐ ☐ ☐ $10$

b $-24$ ☐ ☐ ☐ ☐ ☐ ☐ ☐ ☐ ☐ $9$

c $-36$ ☐ ☐ ☐ ☐ ☐ ☐ ☐ ☐ ☐ $63$

d $-200$ ☐ ☐ ☐ ☐ ☐ ☐ ☐ ☐ ☐ $75$

e $-36$ ☐ ☐ ☐ ☐ ☐ ☐ ☐ ☐ ☐ $30$

**2** People from overseas often think of Australia as a hot country. However, it may surprise you to learn that, as well as recording temperatures of over 40°C, every state and territory in Australia has had below freezing temperatures!

Use the information in the table to complete the tasks below.

| State or territory | Highest temp. | Year and place | Lowest temp. | Year and place |
|---|---|---|---|---|
| ACT | 42.2°C | 1968 (Canberra) | −10°C | 1971 (Canberra) |
| NSW | 50°C | 1939 (Wilcannia) | −23°C | 1994 (Charlotte Pass) |
| NT | 48.3°C | 1960 (Finke) | −7.5°C | 1976 (Alice Springs) |
| QLD | 49.5°C | 1972 (Birdsville) | −11°C | 1895 (Stanthorpe) |
| SA | 50.7°C | 1960 (Oodnadatta) | −8.2°C | 1976 (Yongala) |
| TAS | 40.8°C | 1976 (Hobart) | −13°C | 1983 (Shannon) |
| VIC | 47.2°C | 1939 (Mildura) | −12.8°C | 1947 (Mount Hotham) |
| WA | 50.5°C | 1998 (Mardie) | −6.7°C | 1969 (Booylgoo Spring) |

a In which state or territory was there a difference of 60.5° between the lowest and highest temperatures? _____

b If the lowest recorded temperature in the list were 8° warmer, what would the temperature be? _____

c What is the difference between the lowest and highest temperatures in the Northern Territory? _____

d In which state is there the largest temperature difference between the lowest and highest temperatures, and what is that difference? _____

Oxford University Press

**1**   **a**   Fill in the gaps on this part of an imaginary bank statement.

| Date | Credit $ (Money going in) | Debit $ (Money going out) | Balance $ |
|---|---|---|---|
| 1 June | 500.00 | | 500.00 |
| 3 June | | 125.50 | |
| 5 June | | | 305.50 |
| 10 June | | 420.10 | |
| 16 June | | 81.75 | |
| 21 June | 325.00 | | |
| 30 June | | | −220.30 |

**b**   If $500 were credited on 1 July, what would the new balance be? _____

---

**2**   Here are some rules for a board game called *Opposites*.

- Players start at zero.
- Player A tosses a coin marked *Positive* and *Negative* and then rolls a dice.
- Player A moves forwards (**+**) or backwards (**−**) along a number line according to the number on the dice.
- Player B moves a number of places that is **opposite** to Player A's number.
- After five turns, the player with the higher number wins.

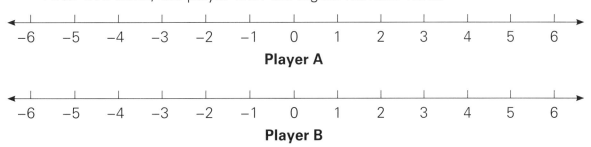

**Player A**

**Player B**

The table below shows the results of the five throws.

**a**   Complete the table to show the number each player moves using + and − to show the direction.

**b**   Plot the players' moves on the number lines.

**c**   Who is the winner and by how much? _____

| Throw | + or − | Number on dice | Player A moves | Player B moves |
|---|---|---|---|---|
| 1 | negative | 2 | −2 | |
| 2 | positive | 3 | | |
| 3 | negative | 4 | | |
| 4 | negative | 1 | | |
| 5 | positive | 6 | | |

## Practice

This fraction wall can help you to find some equivalent fractions.

1. What fractions on the fraction wall are equivalent to $\frac{1}{3}$? _____

2. Find a fraction that is equivalent to each of the following:

   a  $\frac{4}{10}$ _____

   b  $\frac{4}{6}$ _____

   c  $\frac{6}{12}$ _____

   d  $\frac{3}{9}$ _____

   e  $\frac{2}{10}$ _____

   f  $\frac{9}{12}$ _____

3. Fill in the blanks.

   a  $\frac{4}{12} = \frac{\square}{3}$

   b  $\frac{\square}{6} = \frac{2}{3}$

   c  $\frac{8}{\square} = \frac{2}{3}$

   d  $\frac{6}{9} = \frac{2}{\square}$

   e  $\frac{2}{12} = \frac{\square}{6}$

   f  $\frac{\square}{4} = \frac{9}{12}$

Oxford University Press

# Challenge

**1** Divide and shade the shapes to show that:

    **a**   $\frac{2}{3}$ is equivalent to $\frac{4}{6}$                **b**   $\frac{3}{4}$ is equivalent to $\frac{9}{12}$

---

**2**

    **a**   Write two fractions that describe the position of the circle. _____

    **b**   The hexagon is $\frac{9}{12}$ of the way along the line. Write two equivalent fractions that could also describe its position. _____

    **c**   Use equivalent fractions to describe the fraction of the line that is between the circle and the pentagon. _____

    **d**   Draw a star $\frac{20}{24}$ of the way along the line.

    **e**   Write equivalent fractions for the position of the star that you drew in question 2d. _____

---

**3** Fill in all the blanks below to show the equivalent fractions. Show the connection between the numerator and the denominator in each pair of fractions, as shown in question 3a.

    **a**     **b**     **c**     **d**     **e**

    **f**     **g**     **h**     **i**     **j**

1   Reduce these fractions to their lowest equivalent form.

a   $\dfrac{3}{15}$ _____

b   $\dfrac{5}{20}$ _____

c   $\dfrac{18}{24}$ _____

d   $\dfrac{18}{36}$ _____

e   $\dfrac{25}{100}$ _____

f   $\dfrac{45}{50}$ _____

2   The shading on the circles below shows how much pizza is left over after a party.

a   Write the fraction of the total number of slices that are left in its lowest equivalent form. _____

b   Write the amount that has been eaten as a fraction of the total number of slices in as many different ways as you can.

_____

3   A builder pays $408 for a pack of 1200 bricks. 800 of them are found to be broken. The builder telephones to complain that he will have to throw away three-quarters of the bricks and asks for a refund of $306.

a   He is wrong about the fraction. Explain why. _____

_____

b   How much should the refund be in total? _____

c   If three-quarters of the bricks had been broken, how many broken and unbroken

bricks would there have been? _____

4   Imagine you are at a party where there are six guests including yourself. There is a big, rectangular cake.

**A quarter** of the cake must be saved for the rest of the family. The six guests can share what is left. What fraction of the original cake would you each get?

Oxford University Press

### Practice

**1**  a  $\frac{3}{10} + \frac{4}{10} =$ _____  b  $\frac{5}{8} + \frac{2}{8} =$ _____  c  $\frac{1}{5} + \frac{2}{5} =$ _____

   d  $\frac{3}{7} + \frac{3}{7} =$ _____  e  $\frac{7}{10} + \frac{2}{10} =$ _____  f  $\frac{2}{5} + \frac{2}{5} =$ _____

   g  $\frac{1}{8} + \frac{4}{8} =$ _____  h  $\frac{4}{10} + \frac{6}{10} =$ _____  i  $\frac{3}{8} + \frac{5}{8} =$ _____

**2**  a  $\frac{5}{8} - \frac{3}{8} =$ _____  b  $\frac{7}{10} - \frac{3}{10} =$ _____  c  $\frac{9}{12} - \frac{5}{12} =$ _____

   d  $\frac{7}{8} - \frac{2}{8} =$ _____  e  $\frac{3}{5} - \frac{1}{5} =$ _____  f  $\frac{7}{12} - \frac{5}{12} =$ _____

   g  $\frac{4}{7} - \frac{4}{7} =$ _____  h  $\frac{18}{20} - \frac{5}{20} =$ _____  i  $\frac{14}{15} - \frac{4}{15} =$ _____

**3**  a  $\frac{1}{8} + \frac{1}{4} =$ _____  b  $\frac{1}{6} + \frac{2}{3} =$ _____

   c  $\frac{1}{5} + \frac{3}{10} =$ _____  d  $\frac{1}{4} + \frac{5}{8} =$ _____

   e  $\frac{4}{9} + \frac{1}{3} =$ _____  f  $\frac{3}{12} + \frac{1}{6} =$ _____

> Remember:
> Find a common
> denominator
> when adding
> and subtracting
> unlike fractions.

**4**  a  $\frac{7}{10} - \frac{1}{5} =$ _____  b  $\frac{7}{10} - \frac{1}{2} =$ _____  c  $\frac{7}{12} - \frac{1}{2} =$ _____

   d  $\frac{9}{10} - \frac{1}{5} =$ _____  e  $\frac{3}{4} - \frac{1}{12} =$ _____  f  $\frac{3}{10} - \frac{1}{5} =$ _____

**5**  Shade the diagram to solve the addition problem. Write a number sentence that matches the problem.

## Challenge

**1**  Use improper fractions and mixed numbers if the answer is greater than one whole.

a  $\frac{7}{10} + \frac{5}{10} =$ _____  b  $\frac{5}{12} + \frac{11}{12} =$ _____

Here's an example: $\frac{5}{8} + \frac{7}{8} = \frac{12}{8}$. That makes $1\frac{4}{8}$. We can write that as $1\frac{1}{2}$.

c  $\frac{8}{9} + \frac{2}{9} =$ _____  d  $\frac{3}{5} + \frac{3}{5} =$ _____

e  $\frac{7}{9} + \frac{7}{9} =$ _____  f  $\frac{5}{8} + \frac{5}{8} =$ _____

g  $\frac{6}{10} + \frac{7}{10} =$ _____  h  $\frac{1}{3} + \frac{2}{3} =$ _____

**2**  Solve the following:

a  $1\frac{3}{8} - \frac{1}{8} =$ _____  b  $1\frac{4}{9} - \frac{2}{9} =$ _____  c  $1\frac{7}{8} - \frac{4}{8} =$ _____

d  $2\frac{3}{4} - 1\frac{1}{4} =$ _____  e  $1\frac{5}{8} - \frac{5}{8} =$ _____  f  $3\frac{8}{10} - \frac{9}{10} =$ _____

g  $3\frac{2}{9} - \frac{5}{9} =$ _____  h  $3\frac{7}{10} - 1\frac{8}{10} =$ _____

**3**  Find a common denominator to solve the following:

a  $\frac{3}{8} + \frac{3}{4} =$ _____  b  $\frac{2}{5} + \frac{9}{10} =$ _____

c  $1\frac{3}{4} + \frac{5}{8} =$ _____  d  $2\frac{2}{3} + \frac{5}{6} =$ _____

e  $3\frac{7}{8} + \frac{3}{4} =$ _____  f  $2\frac{5}{6} + \frac{1}{3} =$ _____

**4**  Remember to find a common denominator.

a  $1\frac{1}{8} - \frac{1}{4} =$ _____  b  $3\frac{7}{10} - \frac{1}{2} =$ _____

c  $2\frac{3}{12} - \frac{1}{2} =$ _____  d  $1\frac{1}{3} - \frac{5}{6} =$ _____

e  $3\frac{1}{4} - 2\frac{1}{2} =$ _____  f  $1\frac{5}{12} - \frac{2}{3} =$ _____

Oxford University Press

## Mastery

**1** Use the space provided to prove that $\frac{15}{20}$ is **not** equivalent to $\frac{3}{5}$.

**2** Reduce the fraction in the table below to its simplest form by following these instructions:

- Write all the factors of the numerator and of the denominator. (The factors are the numbers that they can be divided by.)
- Circle or underline the **highest common factor** (the highest factor that they can both be divided by).
- Simplify the fraction by dividing the numerator and the denominator by the **highest common factor**.

| Fraction: $\frac{12}{30}$ | Factors of the numerator: |
| | Factors of the denominator: |

The simplest form of $\frac{12}{30}$ is: _____

**3** Simplify the following by finding the **highest common factor** of the numerator and of the denominator and following the procedure in question 2.

a

| Fraction: $\frac{18}{48}$ | Factors of the numerator: |
| | Factors of the denominator: |

The simplest form of $\frac{18}{48}$ is: _____

b

| Fraction: $\frac{25}{45}$ | Factors of the numerator: |
| | Factors of the denominator: |

The simplest form of $\frac{25}{45}$ is: _____

c

| Fraction: $\frac{28}{44}$ | Factors of the numerator: |
| | Factors of the denominator: |

The simplest form of $\frac{28}{44}$ is: _____

### Practice

**1** Shade the following hundred grids to match the decimals and fractions.

a   0.02      b   0.25      c   0.7      d   $\frac{45}{100}$

**2** Write *True* or *False* next to each of the following:

a   0.03 < 0.3 _____

Remember:
> means "is greater than" and < means "is less than".

b   $\frac{1}{1000}$ = 0.001 _____

c   $\frac{99}{1000}$ = 0.99 _____     d   0.03 > $\frac{3}{1000}$ _____

e   $\frac{735}{1000}$ < 0.735 _____     f   $\frac{3}{4}$ > 0.075 _____

g   1.35 = 1$\frac{35}{100}$ _____     h   4.350 > 4.35 _____

**3** Colour the hundred grid as follows:

a   Colour 0.2 blue.

b   Colour $\frac{10}{100}$ green.

c   Colour $\frac{50}{1000}$ yellow.

d   Colour $\frac{1}{4}$ red.

e   Colour 0.39 purple.

f   Write the unshaded amount as a fraction and as a decimal. _____

**4** Order these from **smallest** to **largest**:

| 0.137 | 0.31 | 1.037 | 1.37 | 0.37 | 0.371 |

_____   _____   _____   _____   _____   _____

## Challenge

**1** Complete this table. Write in the missing fractions and decimals.

| | Fraction | Decimal |
|---|---|---|
| a | $\frac{9}{1000}$ | |
| b | | 0.04 |
| c | | 0.001 |
| d | $\frac{16}{1000}$ | |
| e | | 0.25 |
| f | $\frac{750}{1000}$ | |
| g | | 0.099 |
| h | $\frac{1}{2}$ | |

**2** Change the improper fractions in this table to mixed numbers and then to decimals.

| | Improper fraction | Mixed number | Decimal |
|---|---|---|---|
| a | $\frac{5}{4}$ | | |
| b | $\frac{24}{10}$ | | |
| c | $\frac{175}{100}$ | | |
| d | $\frac{250}{100}$ | | |
| e | $\frac{475}{100}$ | | |
| f | $\frac{2750}{1000}$ | | |

**3** Change the following fractions to decimals.

a $\frac{2}{5}$ _____

b $\frac{4}{5}$ _____

c $\frac{5}{20}$ _____

d $\frac{15}{20}$ _____

e $\frac{7}{5}$ _____

f $\frac{25}{20}$ _____

## Mastery

**1** Change the following fractions to decimals. Round them to a maximum of three decimal places.

a $\frac{5}{8}$ _____

b $\frac{5}{7}$ _____

c $\frac{1}{7}$ _____

d $\frac{7}{8}$ _____

e $\frac{7}{12}$ _____

f $\frac{3}{11}$ _____

**2** The final two fractions in the last question had repeating (or recurring) decimals, but you rounded them to three decimal places. When there is one digit that repeats, we can place a dot (or a horizontal line) over the digit to show that it recurs: $0.5\dot{3}$ or $0.5\bar{3}$. If two digits repeat, place a dot (or line) over each digit: $0.1\dot{2}\dot{7}$ or $0.1\bar{2}\bar{7}$.

Find the decimal equivalent of:

a $\frac{2}{3}$ _____

b $\frac{1}{6}$ _____

c $\frac{2}{11}$ _____

d $\frac{5}{6}$ _____

e $\frac{7}{11}$ _____

f $\frac{4}{9}$ _____

**3** Investigate to find three fractions that have recurring decimals. Label your fractions *a*, *b* and *c*.

**4** Not all countries use the decimal point as we do. In some countries a 'dot' separates the thousands from the hundreds. That could make things a bit confusing. Imagine ordering 1.375 kg of sausages and being given one thousand three hundred and seventy-five kilograms instead of 1 kg 375 g!

Do some research to find out some of the countries that use a comma to separate the whole numbers and the decimals.

Oxford University Press

## Practice

**1** Calculate the answers to the following:

**a**
```
    1 • 6  5
+   2 • 5  7
_____
      •
```

**b**
```
  4  5 • 8
+ 2  5 • 9
_____
     •
```

**c**
```
  2  5 • 3  9
+ 3  0 • 7  7  5
_____
       •
```

**d**
```
  7  5 • 8
- 2  7 • 7
_____
     •
```

**e**
```
  7 • 1  5
- 5 • 3  8
_____
    •
```

**f**
```
  6  4  4 • 0  7
- 2  6  8 • 7  8
_____
       •
```

**g**
```
  0 • 2  3
- 0 • 1  7
_____
     •
```

**h**
```
  0 • 6  9
+ 5 • 8  8
_____
    •
```

**i**
```
  3  0  0 • 6  2
- 1  6  4 • 8  3
_____
       •
```

**2** Use place value strategies to write algorithms for the following:

**a** 27.3 + 4.5

**b** 3.45 + 14.2

**c** 31.08 + 17.5

**d** 17.36 − 8.55

**e** 16.2 − 7.57

**f** 24.3 − 15.7

**3**

**a** Building A is 27.34 m high. Building B is 56.2 m high. What is the difference in their heights?

**b** These are the masses of three boxes: 2.86 kg, 9.1 kg and 17.065 kg.

Circle the closest estimate to their total mass.

28 kg          29 kg

30 kg          31 kg

## Challenge

**1** Alice can run twice around some cones on the school oval in 19.05 seconds. She does the first lap in 9.47 seconds. How long does the second lap take?

**2** A party barrel holds 3.5 L of juice. Jack pours in 2.225 L from one bottle and $\frac{3}{4}$ L from another. How much more can the barrel hold?

**3** Henry needs four pieces of wood that are 85 cm long, 1.2 m long, 1.07 m long and half a metre long. He has a 3.9 m length of wood that he can cut them from. How much shorter or longer is it than he needs?

**4** How many 135 g servings of rice can be taken from a 1.5 kg bag?

**5** By how much is 3.75 kg lighter than 4225 g? Write your answer in grams and in kilograms with decimal notation.

Oxford University Press

### Getting to 10 (Part one)

The answer to each of the equations below is exactly 10. Write one digit in each gap to complete the equations. Try to find at least five different solutions.

0. _____ + 5. _____ 3 + _____ . 27 = 10

0. _____ + 5. _____ 3 + _____ . 27 = 10

0. _____ + 5. _____ 3 + _____ . 27 = 10

0. _____ + 5. _____ 3 + _____ . 27 = 10

0. _____ + 5. _____ 3 + _____ . 27 = 10

### Getting to 10 (Part two)

The answer to this puzzle is 10, but you are not given any digits to work with. The rules of the puzzle are:

- One different digit takes the place of a, b, c and d.
- The numbers are read horizontally and vertically as decimals: a.b, c.d, a.c and b.d.
- The sum of the four numbers (a.b + c.d + a.c + b.d) must equal exactly 10.

If we try the digits 6, 1, 3 and 4, the sum becomes 6.1 + 3.4 + 6.3 + 1.4. The total is clearly greater than 10, so other digits must be used. Try to find digits that will solve the problem. If you manage to find a solution, try another. There are quite a few ways to succeed. Happy hunting!

# Multiplication and division of decimals

## Practice

Remember: Put the decimal point in the correct place in the answer box.

**1** Multiply the decimals. Look for patterns in the answers.

a
```
      4 • 1
×         3
```

b
```
    0 • 2  6
×         9
```

c
```
      6 • 9
×         5
```

d
```
    6 • 1  7
×         2
```

e
```
  0 • 4  6  9
×          5
```

f
```
  0 • 5  7  6
×          6
```

g
```
  7  6 • 1  3
×          6
```

h
```
  2  9 • 2  1  8
×             3
```

i
```
  1  9 • 7  5  3
×             5
```

**2** Divide the decimals. Look for patterns in the answers.

a  4 ) 4 8 . 4

b  3 ) 6 . 9 3

c  5 ) 1 7 1 . 5

d  6 ) 2 7 . 2 4

e  6 ) 3 3 . 9

f  7 ) 4 7 . 3 2

g  8 ) 6 2 . 9 6

h  5 ) 4 4 . 9

i  7 ) 6 3 . 7 6 3

j  4 ) 4 9 . 2 8 4

k  6 ) 1 4 0 5 . 9 2

l  3 ) 1 0 3 . 6 2 9

**3** A prize of $4734 is shared by 5 people. How much less than $1000 does each person receive? _____

Oxford University Press

## Challenge

**1** Look for patterns of digits in these multiplications.

**a** 1212.12 × 7

**b** 2323.23 × 7

**c** 34.3434 × 9

**d** 51.5151 × 6

**e** 4545.45 × 5

**f** 565.656 × 9

**g** 37.0037 × 3

**h** 37.0037 × 9

**i** 37.0037 × 6

---

**2** Find the cost of **each individual item** below. Round the answers to an appropriate amount of money.

**a** Three shirts for $39.99 _____

**b** Four games for $73.99 _____

**c** Six chairs for $129.99 _____

**d** Eight books for $99.99 _____

**e** Nine balls for $23.99 _____

**f** Seven balls for $19.95 _____

**g** Five hats for $47.99 _____

**h** Six pictures for $199.95 _____

**1** Rafferty has $145. He wants to buy some cricket balls. They cost $29.35 each. How many can he afford to buy? Show your working out.

## Making one whole

**2** The following nine numbers include whole numbers and decimals:

0.1666   0.3333   0.5   0.6666   1   1.5   2   3   6

It is possible to multiply them in sets of three so that the answer is always 1. The challenge is to arrange them on a 3 × 3 grid so that every row, every column and both diagonals have a product of 1.

They could be placed on the grid as shown in the example on the right.

| 1.5 | 3 | 2 |
|---|---|---|
| 1 | 0.1666 | 0.5 |
| 0.6666 | 0.3333 | 6 |

However, if we multiply the top row, we can easily see that the product is not 1 (1.5 × 3 × 2 = 9), so they must be arranged in a different way. Use the grids below to try to find a solution.
(Hint: If we use the numbers 0.5, 6 and 0.3333, the answer to 0.5 × 6 = 3. And 3 × 0.3333 = 0.9999, which can be appropriately rounded to 1.)

### Practice

Remember: Only use a zero when it is necessary.

Solve these multiplication problems.

**1**
a   4.7 × 10 = _____

b   4.7 × 100 = _____

c   4.7 × 1000 = _____

**2**
a   9.3 × 10 = _____

b   9.3 × 100 = _____

c   9.3 × 1000 = _____

**3**
a   2.39 × 10 = _____

b   2.39 × 100 = _____

c   2.39 × 1000 = _____

**4**
a   3.001 × 10 = _____

b   3.001 × 100 = _____

c   3.001 × 1000 = _____

Dividing by **10** moves each digit **one** place smaller. Dividing by **100** moves each digit **two** places smaller and by **1000** moves each digit **three** places smaller.

Solve these division problems:

**5**
a   234.2 ÷ 10 = _____

b   234.2 ÷ 100 = _____

c   234.2 ÷ 1000 = _____

**6**
a   423 ÷ 10 = _____

b   423 ÷ 100 = _____

c   423 ÷ 1000 = _____

**7**
a   7234.8 ÷ 10 = _____

b   7234.8 ÷ 100 = _____

c   7234.8 ÷ 1000 = _____

**8**
a   260.5 ÷ 10 = _____

b   260.5 ÷ 100 = _____

c   260.5 ÷ 1000 = _____

Solve these multiplication and division problems.

**9**
a   $2.35 × 100 = _____

b   0.003 tonnes × 1000 = _____

c   1.05 m × 10 = _____

**10**
a   $1750 ÷ 1000= _____

b   34 tonnes ÷ 10 = _____

c   524.2 km ÷ 100 = _____

## Challenge

**1** In Unit 1, Topic 7 of the Student Book you looked at multiplying by powers of 10. For example, if you want to calculate 25 × 30, you can split 30 into 3 tens: **Multiply by 10 first**: 25 × 10 = 250. **Then multiply by 3**: 250 × 3 = 750. The same process works for multiplying decimals by powers of 10. For example, if you want to calculate 2.5 × 30, you can split 30 into 3 tens: **Multiply by 10 first**: 2.5 × 10 = 25. **Then multiply by 3**: 25 × 3 = 75

| | × 30 | First × 10 | Then × 3 | Multiplication fact |
|---|---|---|---|---|
| e.g. | 2.5 | 25 | 75 | 2.5 × 30 = 75 |
| a | 1.2 | | | |
| b | 2.3 | | | |
| c | 3.1 | | | |
| d | 1.5 | | | |
| e | 0.32 | | | |

**2** Sometimes it's easier to split the multiple of 10 differently. 2.5 × 30 is the same as finding 2.5 × 3 ten times. **Multiply by 3 first**: 2.5 × 3 = 7.5. **Then multiply by 10**: 7.5 × 10 = 75

| | × 30 | First × 3 | Then × 10 | Multiplication fact |
|---|---|---|---|---|
| e.g. | 2.5 | 7.5 | 75 | 2.5 × 30 = 75 |
| a | 1.3 | | | |
| b | 2.2 | | | |
| c | 3.2 | | | |
| d | 5.1 | | | |
| e | 0.33 | | | |

**3** Use your choice of strategy to multiply these decimals by multiples of 10.

a   1.7 × 40 = _____   b   2.5 × 40 = _____

c   3.5 × 50 = _____   d   3.4 × 30 = _____

e   1.6 × 30 = _____   f   12.5 × 20 = _____

g   1.5 × 50 = _____   h   7.2 × 20 = _____

i   1.7 × 30 = _____   j   $1.20 × 50 = _____

k   $2.25 × 30 = _____   l   2.1 L × 40 = _____

m   8.3 kg × 30 = _____   n   24.8 cm × 60 = _____

**4** A factory sends out 40 packages. Each one has a mass of 1.275 kg. What is the total mass of the packages? _____

Oxford University Press

# Mastery

In Unit 1, Topic 8 you looked at multiplying a whole number by two digits. Here is a reminder of how you can show that 21 × 15 = 315:

```
      2  1
  ×   1  5
      1  0  5  ← 21 × 5
  +   2  1  0  ← 21 × 10
      3  1  5
```

The process is the same for multiplying decimals by two digits. Here is how you can show that 2.1 × 15 = 31.5:

```
      2 • 1
  ×   1   5
      1  0 • 5  ← 2•1 × 5
  +   2  1 • 0  ← 2•1 × 10
      3  1 • 5
```

**1** Multiply:

a
```
      3 • 9
  ×   1   8
      •
  +   •
      •
```

b
```
      2 • 3
  ×   2   5
      •
  +   •
      •
```

c
```
      3 • 3
  ×   2   3
      •
  +   •
      •
```

d
```
      7 • 5
  ×   2   5
  +
```

e
```
      8 • 4
  ×   2   7
  +
```

f
```
      4 • 7
  ×   3   5
  +
```

g
```
      3 • 9
  ×   2   8
  +
```

h
```
      6 • 4
  ×   3   7
  +
```

i
```
      3 • 7
  ×   3   7
  +
```

**2** Nancy is saving for a car. She saves $125.65 each week for a whole year. The car she wants is $6699. How much more or less than the cost of the car does she have by the end of the year?

## Practice

10% has the same value as 0.1 and $\frac{1}{10}$.

1  Shade the grids and fill in the gaps:

a

| Fraction | |
|---|---|
| $\frac{4}{100}$ | |
| Decimal | |
| | |
| Percentage | |
| | |

b

| Fraction | |
|---|---|
| | |
| Decimal | |
| 0.23 | |
| Percentage | |
| | |

c

| Fraction | |
|---|---|
| | |
| Decimal | |
| | |
| Percentage | |
| 45% | |

d

| Fraction | |
|---|---|
| $\frac{4}{10}$ | |
| Decimal | |
| | |
| Percentage | |
| | |

2  Complete the following table:

| | Fraction | Decimal | Percentage |
|---|---|---|---|
| a | | 0.85 | |
| b | | | 3% |
| c | $\frac{7}{100}$ | | |
| d | | | 70% |
| e | $\frac{3}{10}$ | | |
| f | | 0.65 | |
| g | $\frac{3}{4}$ | | |
| h | | | 99% |
| i | $\frac{9}{10}$ | | |
| j | | 0.25 | |
| k | $\frac{3}{5}$ | | |
| l | | | 18% |

3  Write *True* or *False* next to these:

a  $\frac{1}{2}$ = 50% _____

b  0.07 < 70% _____

c  0.23 = $\frac{23}{100}$ _____

d  75% > $\frac{3}{4}$ _____

e  $\frac{4}{100}$ < 40% _____

f  0.89 = 89% _____

g  $\frac{1}{5}$ > 20% _____

h  35% = 0.35 _____

i  1% = 0.01 _____

4  Order these from **smallest** to **largest**:

a  0.7  65%  $\frac{3}{4}$ _____

b  $\frac{3}{10}$  0.292  29% _____

# Challenge

**1** Find the matching fractions, decimals and percentages. Lightly shade each matching set of three a different colour. There is one number that has no match. Put a cross through it.

**2** Fill in the blanks according to their position on this number line. If an arrow does not point to an exact decimal on the line, estimate its position. The first blank has been filled in for you—it is 5% of the way along the number line.

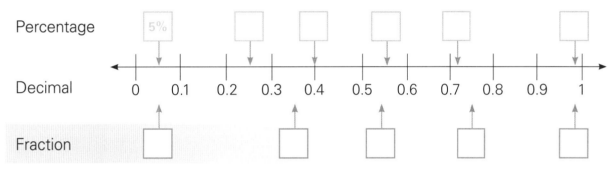

**3** The triangle is 25% of the way along the number line. Choose the best estimate for the positions of the following shapes:

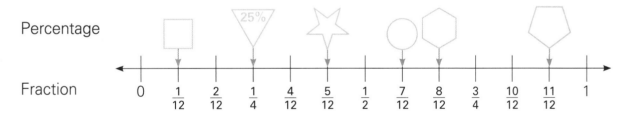

 **a** Square: 9%, 19%, 90% _____  **b** Star: 40%, 50%, 55% _____

 **c** Circle: 48%, 58%, 68% _____  **d** Hexagon: 56%, 66%, 76% _____

 **e** Pentagon: 81%, 91%, 99% _____

**4** Find the decimal equivalents of the positions of the five blank shapes on the number line in question 3 (rounded to two decimal places). Convert the decimals to percentages.

Australia is a very dry place. However, in addition to around one and a half million square kilometres of bushland, there are almost 70 000 km² of lakes and rivers in Australia.

The table below shows the size of some of the world's largest countries. It also shows the area that is covered by forest, and the area of inland waterways in each country.

| | Land area (to nearest $\frac{1}{2}$ million km²) | Area which is forest (to nearest $\frac{1}{2}$ million km²) | Area which is forest (to nearest 1%) | Area of inland water (to nearest 5000 km²) | Area which is inland water (to nearest 0.1%) |
|---|---|---|---|---|---|
| Russia | 17 million km² | 8 million km² | | | 0.5% |
| Brazil | 8.5 million km² | 5 million km² | | | 0.7% |
| Canada | 10 million km² | 3 million km² | | | 9% |
| USA | 10 million km² | | 30% | 660 000 km² | |
| China | 9.5 million km² | | 21% | 270 000 km² | |
| Australia | 7.5 million km² | | 20% | 70 000 km² | |

**1** From the table we can see that almost half of Russia is covered by forest. The actual fraction can be found by dividing 8 million by 17 million. Multiplying the answer by 100 will give the percentage.

    **a** Fill in the percentage of forest area for Russia, Brazil and Canada. Round to the nearest 1%.

    **b** Fill in the area of forest in the table for USA, China and Australia. Round to the nearest half million km².

    **c** Fill in the percentage of inland water area in the table for USA, China and Australia. Round to the nearest 0.1%.

    **d** Fill in the area of inland water for Russia, Brazil and Canada. Round to the nearest 5000 km².

**2** The figures in the table above are all rounded. Research the actual figures for one or two countries. You could begin with Australia. Find out how close the actual figures are to the figures in the table. Comment on how important it is to have true figures when showing information like this.

Oxford University Press

**Practice**

1  Calculate the new price of each ball.

a

Normally: $49.99
Discount: 10%

New price: _____

b

Normally: $19.99
Discount: 25%

New price: _____

c

Normally: $15.99
Discount: 10%

New price: _____

d

Normally: $9.99
Discount: 25%

New price: _____

e

Normally: $39.99
Discount: 10%

New price: _____

f

Normally: $3.99
Discount: 10%

New price: _____

g

Normally: $69.99
Discount: 10%

New price: _____

h

Normally: $4.99
Discount: 25%

New price: _____

i

Normally: $14.99
Discount: 10%

New price: _____

j

Normally: $24.99
Discount: 50%

New price: _____

## Challenge

Sometimes shops have discounts of 20% and 15%.

If a book has a discount of 20%, it's like two lots of 10%.

10% of $10 is $1. So, 20% of $10 is $2. The new price is $8.

If the book has a discount of 15%, it's like a discount of 10% plus half of 10%. 10% of $10 is $1. Half of $1 is $0.50. So, 15% of $10 is $1 + $0.50 = $1.50. The new price is $8.50.

Normal price: $10

**1** Calculate the discount and write the new price of the following.

**a**

$234.99
Less 25%

New price: _____

**b**

$117.45
Less 10%

New price: _____

**c**

$97.95
Less 10%

New price: _____

**d**

$79.99
Less 20%

New price: _____

**e**

$229.99
Less 20%

New price: _____

**f**

$984.99
Less 20%

New price: _____

**g**

$19.99
Less 15%

New price: _____

**h**

$119.99
Less 20%

New price: _____

**i**

$179.99
Less 15%

New price: _____

**j**

$549.99
Less 5%

New price: _____

**1** Most calculators have a key to help work out percentages quickly. Use a calculator for the following activities.

Here are five sequences of calculator keystrokes. Entering **two** of these into the calculator exactly as written will give the answer that is equal to 10% of $135. Circle them.

a   10, x, %, 135, =      b   %, 10, x, 135, =

c   10, %, x, 135, =      d   135, 10, x, %, =

e   135, x, 10, %, =

*Calculate the answer in your head first so that you know whether the calculator is giving the correct answer.*

**2** When someone borrows money, the interest rate is rarely an easy, rounded percentage. Percentages such as 9.5% are quite common.

Finding 9.5% of an amount such as $15.99 is easy with a calculator. However, it is useful to estimate the amount so that we know whether to trust the calculator answer. 9.5% can be rounded to 10%, so the answer must be close to 10% of $16, which is $1.60.

Round the figures to estimate the answers for the following. Then use a calculator method of your choice to work out the exact answer. Be sure to round the calculator answer to the nearest cent.

a   9.5% of $16        Estimate: $1.60        Exact answer: _____

b   4.9% of $30        Estimate: _____      Exact answer: _____

c   9.9% of $75        Estimate: _____      Exact answer: _____

d   19.5% of $80       Estimate: _____      Exact answer: _____

e   5.19% of $20       Estimate: _____      Exact answer: _____

f   15.35% of $60      Estimate: _____      Exact answer: _____

g   9.7% of $7500      Estimate: _____      Exact answer: _____

h   10.25% of $5999    Estimate: _____      Exact answer: _____

**3** Imagine someone borrowed $15 999 to buy a car at an interest rate of 5.16% per annum (year). If they paid the money back by the end of the year, how much interest would they have paid?

## Practice

**1** Write a rule for this pattern of sticks.

_____

**2** **a** Look at the pattern above and complete the following table by working out how many sticks you would need to make the number of pentagons specified.

| Number of pentagons | 1 | 2 | 3 | 4 | 5 | 6 | 7 | 8 | 9 | 10 |
|---|---|---|---|---|---|---|---|---|---|---|
| Number of sticks | 5 | | | | | | | | | |

**b** How many sticks would be needed for a pattern of 20 pentagons? _____

**3** Look at the way this pattern has been made.

A rule for making the pattern could be:

_Use five sticks for the first pentagon and then four for every other pentagon._

Which of these rules could also be used to describe the way pattern is made? (Circle one.)

- Use five sticks for every pentagon.
- Use five sticks for the first pentagon, four for the second pentagon, three for the third pentagon, two for the fourth pentagon and one less for every other pentagon.
- Start with one stick and then use four sticks for every pentagon.

**4** Follow this rule for the patterns of pentagons and then write a number sentence to show how many sticks are needed for each pattern.

| Pattern | Rule | How many sticks are needed? |
|---|---|---|
| **a** | Use five sticks for the first pentagon and then four for every other pentagon. | $5 + 4 \times \boxed{\phantom{0}} = \boxed{\phantom{0}}$ |
| **b** | | $5 + \boxed{\phantom{0}} = \boxed{\phantom{0}}$ $5 + \boxed{\phantom{0}} \times \boxed{\phantom{0}} = \boxed{\phantom{0}}$ |

**5** Write a rule for this pattern.

_____

# Challenge

**1** This flow chart shows a rule of divisibility by 9.
It can be used for testing whether any number is
divisible by 9.

Use the flow chart to decide whether or not these
numbers are divisible by 9. Answer *Yes* or *No*.
Prove that the rule works for any number by doing
a division that shows whether or not there is a
remainder.

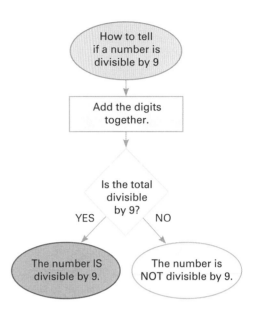

**a** 972 _____

**b** 799 _____

**c** 441 _____

**d** 4284 _____

**e** 3849 _____

**f** 123 723 _____

**2** Use the rule of divisibility by 9 to write a number with six
different digits that you know is divisible by 9. Prove the
rule by writing a division problem.

**3** On a separate piece of paper or with a computer program, write two flow charts: one
that will show whether a number is divisible by 5 and the other for divisibility by 6.

**a**

**b**

**4** This function machine is called FIDO.
It can be programmed to do the same
thing to any number that is fed into it.

   **a** Work out what this FIDO machine
   does to the numbers fed into it.

   Write each of the two operations
   inside the ovals on the machine.

   **b** What would be drawn out if 1.5
   were fed into the FIDO machine?

_____

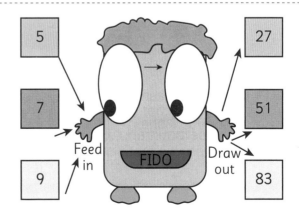

**1** You will probably recognise the numbers 2, 3, 5, 7 and 9 as the single-digit prime numbers. But what about 2-digit numbers? How can you work out whether or not they are prime numbers? First, you must ask: *Is the number divisible by 2?*

Design a flow chart that will serve as a test to show whether any 2-digit number is a prime number. It has been started for you below.

A flow chart needs to be neat and easily understood by anyone who reads it.

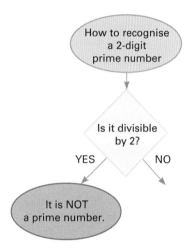

How to recognise a 2-digit prime number

Is it divisible by 2?

YES          NO

It is NOT a prime number.

**2** Use your flow chart to make a list of as many 2-digit prime numbers as you can.

## Practice

**1** Use the order of operations to complete the following:

a $5 \times 2 + 4 = $ _____

b $3 + 8 \times 2 = $ _____

c $24 \div 4 - 3 = $ _____

d $(6 - 3) + (10 - 3) = $ _____

e $2 + 54 \div 9 = $ _____

f $36 \div (8 - 4) = $ _____

g $25 - 12 \div 3 = $ _____

h $(35 - 7) \div 2 = $ _____

i $\frac{1}{2}$ of $9 + 3 = $ _____

j $\frac{1}{2}$ of $(9 + 3) = $ _____

k $3 \times 3^2 = $ _____

l $(3 \times 3)^2 = $ _____

m $(4 + 5)^2 = $ _____

n $4 + 5^2 = $ _____

o $\frac{1}{4}$ of $12 - 2 = $ _____

p $\frac{1}{4}$ of $(12 - 2) = $ _____

**2** Use equations to split the number you are multiplying by.

| | Problem | | Split the problem to make it simpler | | Rewrite the problem | | Answer |
|---|---|---|---|---|---|---|---|
| a | $42 \times 3$ | = | $(40 \times 3) + (2 \times 3)$ | = | $120 + 6$ | = | |
| b | $14 \times 6$ | = | | = | | = | |
| c | $53 \times 5$ | = | | = | | = | |
| d | $23 \times 9$ | = | | = | | = | |

**3** Changing the order of operations.

| | Problem | | Change the order to make it simpler | | Rewrite the problem | | Answer |
|---|---|---|---|---|---|---|---|
| a | $25 \times 11 \times 4$ | = | $25 \times 4 \times 11$ | = | | = | |
| b | $50 \times 12 \times 4$ | = | | = | | = | |
| c | $8 \times 15 \times 4$ | = | | = | | = | |
| d | $5 \times 19 \times 2$ | = | | = | | = | |

## Challenge

**1** Rewrite each equation in the table below using "opposites" to find the value of ◆.

| | Problem | Use opposites | Write the value of ◆ | Check by writing the equation |
|---|---|---|---|---|
| **a** | $◆ × 7 = 63$ | $◆ = 63 ÷ 7$ | | _____ $× 7 = 63$ |
| **b** | $◆ - 2\frac{1}{2} = 6$ | | | |
| **c** | $\frac{1}{4}$ of $◆ = 15$ | | | |
| **d** | $◆ × 100 = 225$ | | | |
| **e** | $◆ ÷ 100 = 1.6$ | | | |
| **f** | $◆ ÷ 8 = \frac{1}{8}$ | | | |
| **g** | $◆ × 1000 = 7225$ | | | |

**2** Substitution: Put a number in place of the ◆ symbol to see if the equation balances. Circle or highlight your choice for the substitute number. Write a number sentence to check.

| | Problem | Possible substitutes for ◆ | | | | Check |
|---|---|---|---|---|---|---|
| **a** | $◆ × (2 + 3) = 50$ | 8 | 9 | 10 | 11 | |
| **b** | $72 ÷ ◆ - 5 = 3$ | 9 | 10 | 11 | 12 | |
| **c** | $1\frac{1}{2} × ◆ + 6 = 10\frac{1}{2}$ | 2 | 3 | 4 | 5 | |
| **d** | $25 ÷ ◆ + 2.5 = 5$ | 5 | 10 | 15 | 20 | |
| **e** | $25 × (10 - ◆) = 150$ | 3 | 4 | 5 | 6 | |
| **f** | $◆ × 2 = 3 + 5^2$ | 12 | 13 | 14 | 15 | |
| **g** | $(5 + ◆) × 10 = 25 × 3$ | 1.5 | 2 | 2.5 | 3 | |

**3** Solve the following by substitution:

**a** $◆ × 4 = 10^2$ _____

**b** $(6 ÷ 4) × ◆ = 3$ _____

**c** $4 + 2 × ◆ = 10$ _____

**d** $\frac{1}{2}$ of $6 × ◆ = 15$ _____

Oxford University Press

### Problem-solving

When solving problems, we usually use the technique that is quickest. One method is to take out all the unnecessary information and write an equation.

**1**  Two classes are going out for sport and need hoops. The sport teacher already knows that they will need 150 hoops, but she gives two students the following problem on a piece of paper:

*There are **27** students in your class and **23** in the other class. We need **three hoops** for **each child**. Please fetch the correct number of hoops.*

They write an equation using just the important information.

The number sentence they write is **27 + 23 × 3**.

However, unless they use the order of operations correctly, they will not get the correct number. Solve both equations, and circle the correct one.

$27 + 23 \times 3 =$ _____          $(27 + 23) \times 3 =$ _____

**2**  Imagine that the story above changes slightly. The teacher also needs one hoop for herself. Rewrite the equation showing the total number of hoops that are needed.

**3**  Write equations that will solve the following problems and use them to find the correct answers.

a  Henry has $4 in his piggy bank and his aunty gives him another $2. His mum says she will double the amount. The same day Henry gets three birthday cards, each with a $5 note inside. The next day he gets five cards: one has $2, two have $5 notes, and two have $10 notes. How much does he have in total?

b  Finlay has a new novel. It has 96 pages. During the first two days he reads five pages a day. The next day, he reads a quarter of the book. He then reads 10 pages a day for two days. How many pages are there left to read by that time?

## Practice

**1** Convert between the units of length.

a

| cm | mm |
|---|---|
| | 20 mm |
| 3.5 cm | |
| | 47 mm |
| 12.3 cm | |
| | 150 mm |

b

| m | cm |
|---|---|
| 4 m | |
| | 250 cm |
| 2.32 m | |
| | 435 cm |
| 7.5 m | |

c

| km | m |
|---|---|
| | 3000 m |
| 14 km | |
| | 4750 m |
| 3.9 km | |
| | 1800 m |

**2** Match **two** lengths to each object in the table below.

2.7 m          2.7 km

75 mm          27 mm          2700 mm

2700 m          127 cm

2.7 cm          1.27 m          7.5 cm

| | Object | 1st unit | 2nd unit |
|---|---|---|---|
| a | The length of a paper clip | | |
| b | The height of a young student | | |
| c | The length of a finger | | |
| d | The height of a ceiling | | |
| e | The length of a street | | |

**3** Measure each line in the table below and write the length three ways.

| | | mm | cm and mm | cm |
|---|---|---|---|---|
| a | ▬▬▬▬▬▬ | | | |
| b | ▬▬▬▬▬▬▬▬▬ | | | |
| c | ▬▬▬▬▬▬▬ | | | |
| d | ▬▬▬▬▬▬▬▬▬▬ | | | |

Oxford University Press

# Challenge

**1** Circle any of the following that do **not** describe the length of a 3.05 m room.

305 cm          3 m 5 cm          3 m 50 cm          3050 mm          350 cm

---

**2** Line B is 8.5 cm long.

**a** Estimate (DO NOT MEASURE) the lengths of the other two lines.

Line A ━━━━━━━━━━━━━━━━━━━━━━

I ine B ━━━━━━━━━━━━━━━━━━━━

Line C ━━━━━━━━━━━━━━━━━━━━━━━

**b** Write the lengths of Lines A and C to the nearest millimetre.

Line A: _____          Line C: _____

---

**3** **a** The perimeter of the triangle is 9 cm. What is the smallest number of sides that you would need to measure to calculate the perimeter? _____

**b** What is the length of each side of the triangle? _____

---

**4** Find the perimeter of each shape below. Write down the number of sides that you measured to calculate the perimeter of each shape.

**a**

Perimeter = _____
Number of sides
I measured: _____

**b**

Perimeter = _____
Number of sides
I measured: _____

---

**5** Draw a rectangle with a perimeter of exactly 34.8 cm. Use the line below as one of the sides of your rectangle. Write the lengths on one long and one short side of the rectangle.

_____

1   This is the Peel P50—the smallest production car ever made. Production started and ended in the 1960s. They are very rare, and can sell at auction for more than $100 000!

The Peel P50 is 1.34 m long, 1.2 m high and only 0.99 m wide.

a   Write your height in cm. _____

b   What is the difference between your height and the height of the P50? _____

c   Write the width of the P50 using a different unit of length. _____

d   What is something in your room that is about the same as the length of the P50 car? _____

2   In 2012 an even smaller car was built by Austin Coulson from Phoenix, Arizona, USA. It is allowed on the roads, but it cannot be called a production car because only one was ever made. The car is 7.53 cm shorter and 56.5 cm lower than the P50. (It has no roof, of course!) It is also an amazing 33.59 cm narrower than the tiny Peel P50.

a   What are the dimensions of Austin Coulson's car? _____

b   With your teacher's permission, search online for "Smallest car, Austin Coulson, 2012" to have a look at the car.

c   What is something in a home or a classroom that is about the size of the smallest car? _____

3   The smallest book in your classroom would probably not be as small as the one in this photo!

The smallest comic book was made in the 1990s by Martin Lodewijk. It is only 37 mm high. Its width is 1.12 cm less than its height. Write the width of the comic book in cm and in mm. _____

4   3D printers can make amazingly small objects—so small that they cannot even be measured in millimetres. They are measured in **micrometres**. The smallest replica guitar in the world was made using a 3D printer at a university in New York. It's a very tiny version of the guitar in this photo, but it is actually smaller than one of the full stops on this page!

The tiny model guitar is only 10 micrometres long! Find out what you can about micrometres so that you can appreciate how amazing the replica guitar is.

Oxford University Press

## Practice

1   Measure each rectangle below to calculate its area.

a

Area = _____

b

Area = _____

c

Area = _____

d

Area = _____

e

Area = _____

- - - - - - - - - - - - - - - - - - - - - - - - - - - - - - - - - - - - - - - - - - - - - - - - - -

2   This is a floor plan of a housing unit. Using a scale of 1 cm: 1 m, write the area of each room.

Lounge room

Balcony

Kitchen/ Dining room

Bedroom

Bathroom

Balcony = _____ m²

Bedroom = _____

Bathroom = _____

Lounge room = _____

Kitchen/Dining room = _____

## Challenge

**1** Measure and then calculate the area of each shape.

a

Area = _____

b

Area = _____

c

Area = _____

d

Area = _____

e

Area = _____

**2** Draw two different shapes that are not rectangles. Each shape must have an area of 24 cm².

Oxford University Press

**1** Find the area of each right-angled triangle.

a

b

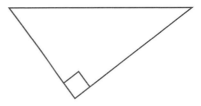

Area = _____

Area = _____

---

**2** Find the area of each triangle.

a

?

Area = _____

b

Area = _____

---

**3** Design a pen for a dog to run around in. You have 36 m of chicken wire and plenty of stakes.

Start with a rectangular enclosure. A pen 17 m long by 1 m wide would use all the wire, but would other dimensions provide a bigger area?

Draft some plans for the dog pen. Find one that provides the biggest area whilst using all 36 m of wire. Using a suitable scale, draw it on a separate piece of paper. (Grid paper might be useful.)

You could also design a non-rectangular shape so that the dog would have places to hide, sleep, feed and play. On your final copy, show the position of the kennel, the play area etc.

## Practice

**1** **a** How many centimetre cubes would be needed to make this model? _____

**b** What is its volume? _____

1 cm
3 cm
6 cm

**2** Find the volume of each of the following objects:

**a**

5 cm
8 cm
2 cm

Volume: _____ cm³

**b**

3 cm
3 cm
7 cm

Volume: _____ cm³

**c**

3 cm
4 cm
7 cm

Volume: _____ cm³

**3** Complete the following tables to convert between the different units of volume and capacity.

**a**

| Kilolitres | Litres |
|---|---|
| 5 kL | |
| 2.5 kL | |
| | 4250 L |
| | 3750 L |

**b**

| Litres | Millilitres |
|---|---|
| 4 L | |
| | 3500 mL |
| 2.25 L | |
| | 9750 mL |

**c**

| Volume | Capacity |
|---|---|
| 100 cm³ | mL |
| | 500 mL |
| 175 cm³ | |
| | 2 L |

**4** Put these capacities in order from smallest to largest:

Orange 1.25 L

Mango 600 mL

Cola 1 L 200 mL

Water 1½ L

Apple 1375 mL

Lemon ¾ L

Water 375 mL

Oxford University Press

## Challenge

**1** Using the capacities of the drinks on page 68, shade the jugs to show the level in the jug after the following drinks have been poured in:

**a** 2 mango drinks

Amount: _____

**b** 1 orange drink

Amount: _____

**c** 1 orange and 1 lemon drink

Amount: _____

**d** 1 large and 1 small water

Amount: _____

**e** 1 apple and 1 mango drink

Amount: _____

**f** 1 apple drink and 1 small water

Amount: _____

**2** Using centimetre cubes, how could you prove that you would need 30 centimetre cubes to make this model?

_____

_____

**3** Use centimetre cubes to make a different rectangular prism from the one in question 2 that also has a volume of 30 cm³. Sketch the rectangular prism and write the length, width and height on your drawing.

**1** Very large quantities of water are measured in megalitres (ML) and gigalitres (GL). Find out what you can about these units of capacity.

**2** Taking as much care as possible, sketch as many **different** rectangular prisms as you can that have a volume of 24 cm$^3$. Write the length, width and height on each model.

## Practice

**1**  Complete the tables to convert between the different units of mass.

a

| Tonnes | Kilograms |
|--------|-----------|
|        | 2000 kg   |
| 3.5 t  |           |
|        | 4250 kg   |
| 5.175 t |          |
|        | 975 kg    |

b

| Kilograms | Grams  |
|-----------|--------|
| 5 kg      |        |
|           | 3500 g |
| 0.75 kg   |        |
|           | 450 g  |
| 3.07 kg   |        |

c

| Grams  | Milligrams |
|--------|------------|
| 3 g    |            |
|        | 2250 mg    |
| 1.5 mg |            |
|        | 735 mg     |
|        | 1 mg       |

**2**  Record the mass of each box below, taking note of the increments on the scales.

a

Mass: _____

b

Mass: _____

c

Mass: _____

**3**  Using the scales in question 2, which would you use if you needed to have:

a  50 g of sugar? _____

b  3.25 kg of fruit? _____

c  $1\frac{3}{4}$ kg of onions? _____

d  1200 g of rice? _____

**4**  Draw a pointer on the scale to show that the box has a mass of 4.4 kg.

**1** Estimate the mass of each box and write it in as many different ways as you can.

**a**

**b**

**c**

**2**

**A**

**B**

**C**

**D**

**a** Order the trucks A, B and C from lightest to heaviest load. _____

**b** Truck D has a load that is greater than the load of Truck B but less than the load of Truck A. Write a possible mass for Truck D's load in tonnes with a decimal; in tonnes and kilograms; and in kilograms. _____

_____

**c** What is the difference between the mass of the loads on trucks B and C?

_____

**3** You will need a set of weighing scales for this task. Find three items in your classroom (not weights) that you think will have a total mass of 1 kg.

**a** Write your estimate for the mass of each in the table, making sure that the total is 1 kg.

**b** Weigh each object and record its mass.

**c** How close to 1 kg was your estimate? _____

| Object | My estimate | Actual mass |
|--------|-------------|-------------|
|        |             |             |
|        |             |             |
|        |             |             |
| **Total** |          |             |

Oxford University Press

1  Match each truck to its load.

**Truck A**

This truck is carrying 2.8 tonnes.

**Truck B**

This truck is carrying 2.525 tonnes.

**Truck C**

This truck is carrying 2.67 tonnes.

a

| | |
|---|---|
| 330 kg | 570 kg |
| 0.62 t | 1.15 t |

This is the load for

Truck: _____

b

| 0.475 t | 1.45 t | 255 kg |
|---|---|---|
| | | 345 kg |

This is the load for

Truck: _____

c

| 0.075 t | |
|---|---|
| 0.4 t | 900 kg |
| 1 t 425 kg | |

This is the load for

Truck: _____

---

2  a  Write the total mass of the 12 large apples in kilograms with a decimal; in kilograms and grams; and in grams.

_____

b  What is the mean average mass of one apple (to the nearest gram)? _____

c  No two apples have exactly the same mass. Write a possible mass for each of the 12 apples.

_____

_____

## Practice

**1** Fill in the missing times.

a

**During the night**

b

c

**am/pm time**    4:34 pm    _____    _____

**24-hour time**    _____    _____    1313

**In the morning**

d

e

f

**am/pm time**    _____    11:47 am    _____

**24-hour time**    _____    _____    0003

Gold rush!

**2** Use the information to complete the timeline. Fill in the years and put the correct letters in the boxes. Draw an arrow from each box to the correct place on the timeline.

- 1853: Mining licence fees reduced to 10 shillings—**L**
- 1861: One-fifth of Victorian adults come from China—**S**
- 1851: Edward Hargraves discovers gold in NSW. Mining licence fee is 30 shillings—**G**
- 1855: 10 000 Chinese miners arrive in Victoria—**R**

- 1858: Gold is discovered in Northern Queensland—**U**
- 1854: 28 men are killed at the Eureka Stockade, Victoria—**D**
- 1852: 100 000 prospectors arrive from overseas—**O**
- 1864: Gold is discovered at Coolgardie, Western Australia—**H**

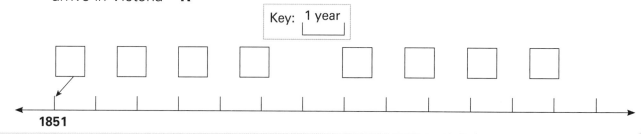

Key: 1 year

1851

Oxford University Press

## Challenge

**1** Use the information in this train timetable to complete the tasks below.

**a** How long does the first train take to get from Springwood to Penrith?

_____

**b** At what time would you leave Blaxland to reach Emu Plains by quarter to seven? _____

| Station | Time | | | | | | | |
|---------|------|------|------|------|------|------|------|------|
| Springwood | 0513 | 0543 | 0613 | 0631 | 0643 | 0701 | 0713 | 0731 |
| Valley Heights | 0516 | 0546 | 0616 | ... | 0646 | ... | 0716 | ... |
| Warrimoo | 0520 | 0550 | 0620 | ... | 0650 | ... | 0720 | ... |
| Blaxland | 0524 | 0554 | 0624 | 0640 | 0654 | 0710 | 0724 | 0740 |
| Glenbrook | 0529 | 0559 | 0629 | 0645 | 0659 | 0715 | 0729 | 0745 |
| Lapstone | 0533 | 0603 | 0633 | ... | 0703 | ... | 0733 | ... |
| Emu Plains | 0540 | 0610 | 0640 | 0655 | 0710 | 0725 | 0740 | 0755 |
| Penrith | 0544 | 0614 | 0644 | 0659 | 0714 | 0729 | 0744 | 0759 |

**c** How long is the journey from Springwood to Blaxland if you catch the train that leaves around half past six? _____

**d** Imagine you want to be at a friend's house in Glenbrook at quarter to eight in the morning. It is a 15-minute walk from Glenbrook station to get there. Which train would you catch from Valley Heights? _____

**e** Which train would you take from Springwood to get to Lapstone by 7 am?

_____

**2** This timetable shows some of the return times from Penrith to Springwood. Use the information from both timetables to complete the tasks below.

**a** At what time does the 1310 train from Penrith arrive at Springwood? Answer in am/pm time.

_____

| Station | Time | | | | | | | |
|---------|------|------|------|------|------|------|------|------|
| Penrith | 1110 | 1210 | 1310 | 1410 | 1510 | 1610 | 1710 | 1839 |
| Emu Plains | 1113 | 1213 | 1313 | 1413 | 1513 | 1613 | 1713 | |
| Lapstone | 1119 | 1219 | 1319 | 1419 | 1519 | 1619 | 1719 | |
| Glenbrook | 1124 | 1224 | 1324 | 1424 | 1524 | 1624 | 1724 | |
| Blaxland | 1129 | 1229 | 1329 | 1429 | 1529 | 1629 | 1729 | |
| Warrimoo | 1132 | 1232 | 1332 | 1432 | 1532 | 1632 | 1732 | |
| Valley Heights | 1136 | 1236 | 1336 | 1436 | 1536 | 1636 | 1736 | |
| Springwood | 1140 | 1240 | 1340 | 1440 | 1540 | 1640 | 1740 | 1902 |

**b** If you took the train at about 7 am from Springwood to Penrith, how long would you spend there if you caught the return train at 10 minutes past noon? _____

**c** What is the difference between the shortest and longest times to get from Penrith to Springwood? _____

**d** What do you notice about the longest journey between Penrith and Springwood and the longest journey in the other direction? _____

_____

### The first people on the moon

**1** In 1969, the Apollo 11 space flight landed the first people on the Moon. All of the clocks used during the mission were set to the time on the east coast of America.

a The list below contains some information from the Apollo 11 mission. The table contains a series of times and dates. Match each event in the list to the corresponding date and time by working out the order in which you think the events must have occurred. Write the letter from the table that corresponds to each event in the space provided.

- Neil Armstrong and Buzz Aldrin's lunar module touched down on the Moon with just a few seconds of fuel left. _____

- Armstrong and Aldrin were relieved when their little spacecraft left the Moon's surface. _____

- The three astronauts splashed down in the Pacific Ocean. _____

- Apollo 11 blasted off from Earth on a four-day journey to the Moon. _____

- Neil Armstrong took a step on the Moon and called it a "giant leap for mankind". _____

- The lunar module separated from the main spacecraft and headed for the Moon's surface. _____

| 9:37 am 16 July | 1:47 pm 20 July | 4:18 pm 20 July | 10:56 pm 20 July | 1:54 pm 21 July | 12:55 pm 24 July |
|---|---|---|---|---|---|
| A | B | C | D | E | F |

b Write the event letters from the table in question 1a on the timeline and, as accurately as possible, draw an arrow to show the date and time of each event. The taller marks represent midnight.

A     Apollo 11, 1969     ☐ ☐ ☐ ☐          ☐

16 July   17 July   18 July   19 July   20 July   21 July   22 July   23 July   24 July

c At what time did Neil Armstrong first step onto the Moon? Answer in 24-hour time. _____

d How many hours and minutes was it from C to D? _____

e How many days, hours and minutes was it from A to B? _____

f How long did the whole trip last? _____

Oxford University Press

## Practice

**1** Write the name of the shapes below. Draw horizontal stripes on the regular shapes and vertical stripes on the irregular shapes.

a

_____

b

_____

c

_____

d

_____

e

_____

f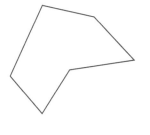

_____

**2** Explain why this shape is not a polygon.

_____

_____

**3** Circle the sentence that best describes this polygon:

It is an irregular quadrilateral.
It is an irregular quadrilateral with one right angle.
It is an irregular quadrilateral with two pairs of parallel sides.
It is an irregular quadrilateral with one pair of parallel sides.
It is an irregular quadrilateral with no parallel sides.

**4** Draw a regular quadrilateral.

## Challenge

**1** Write the name of each type of triangle and write about the properties that helped you to identify it. You may need a ruler and/or a protractor to help you identify them.

| | | Name | Properties |
|---|---|---|---|
| **a** | | | |
| **b** | | | |
| **c** | | | |
| **d** | | | |
| **e** | | | |

**2** Look carefully at the following 2D shapes. For one reason or another, each of them could be described as "the odd one out".

**a**     **b**     **c**     **d**     **e**

Choose one (or more!) of the shapes and explain why you think it is "the odd one out". _____

**3** Draw the shape described below using a pencil and ruler.

- It has four sides.
- Two of the angles are obtuse and two are acute.
- It has two pairs of parallel sides.
- The opposite sides are the same length.
- The base is 9 cm long.

Oxford University Press

**1** In the space below, draw a circle with a radius of 4 cm. Label the following:

- the circumference
- a diameter
- a quadrant
- a sector
- a radius
- a semi-circle.

### Drawing a perfect square with a pair of compasses

**2** It may seem hard to believe, but it is possible to draw a perfect square from a circle! (You will need some plain paper.) Here's how:

**Step 1:** Draw a circle of any size. Draw a diameter and put two small dots (**A** and **B**) on the diameter. (See Step 1 diagram.)

**Step 2:** Set your compass at more than half the distance between **A** and **B**. Put the point of the compass at dot **A** and draw arcs above and below the diameter. Repeat from dot **B**. (See Step 2 diagram.)

**Step 3:** Use a ruler to draw a second diameter that passes through the places where the arcs cross. (See Step 3 diagram.)

**Step 4:** Join the ends of the two diameters to make a perfect square. (See Step 4 diagram.)

You might need a few practice attempts before drawing your final version.

**Step 1**  **Step 2**  **Step 3**  **Step 4**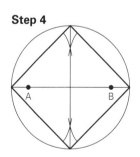

**3** It is possible to draw other polygons with the aid of a pair of compasses. You could start, with your teacher's permission, by searching online for "Drawing a hexagon with a pair of compasses".

### Practice

**1** Name these 3D objects. Say whether each is a *prism*, a *pyramid* or *other 3D object*.

a

_____
_____

b

_____
_____

c

_____
_____

d

_____
_____

e

_____
_____

f

_____
_____

**2** Fill in the gaps in the table.

| | Object | Shape of the base(s) | Number of bases |
|---|---|---|---|
| a | | | |
| b | | | |
| c | | | |
| d | | | |

1  Copy each object onto the grid and then make a freehand sketch in the space to the right.

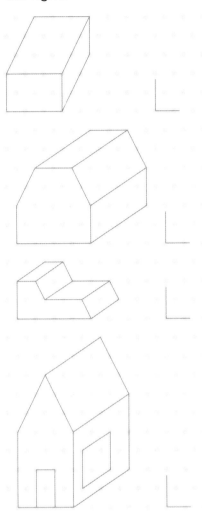

2  These objects below are not drawn on a grid. Decide on a suitable size and then use the isometric grid to sketch at least two copies of each object. You may wish to use spare isometric dot paper for further practice.

### Cross-sections of 3D objects

**1** The arrows in the drawings below show the direction of the cut for each cross-section. In this question the cuts are all parallel to the base. Identify and draw the cross-section of each 3D object.

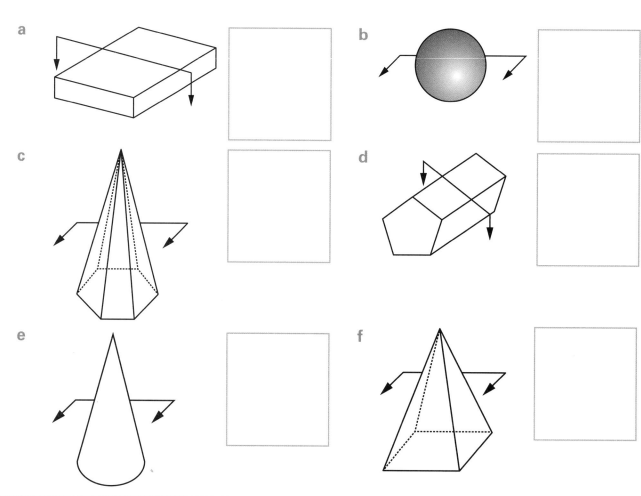

a

b

c

d

e

f

**2** In the following cross-sections the cut is **not** parallel to the base. Draw the **oblique cross-section** for each object.

a

b

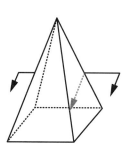

**3** The following could all be **oblique cross-sections** of two 3D objects. Name the objects and explain why.

Oxford University Press

# Angles

## Practice

**1** Write the type of each angle. Write the size of the angle inside the arc.

a

b

c

**2** How do you know that the size of this reflex angle is 330°?

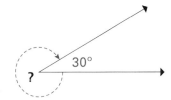

_____

**3** Write the size of each reflex angle.

a

b

c

1  Measure the following angles to the nearest degree.

a  b

2  Draw angles of 37° from the dot on each base line below.

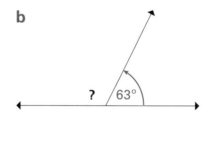

3  Calculate the size of each unknown angle without a protractor.

a

?  95°

b

?  63°

c

?

272°

d

38°

?

e

56°  ?

?

**1** Calculate the size of all the missing angles by measuring **only** the angle marked in red.

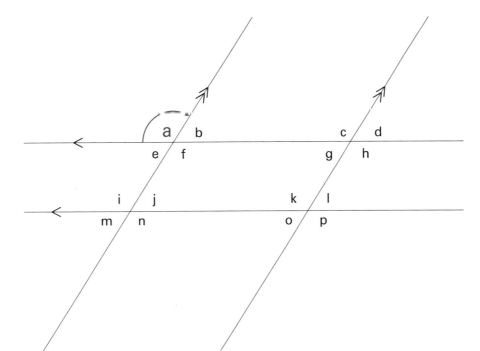

| | | | | | | | |
|---|---|---|---|---|---|---|---|
| a | = _____ | b | = _____ | c | = _____ | d | = _____ |
| e | = _____ | f | = _____ | g | = _____ | h | = _____ |
| i | = _____ | j | = _____ | k | = _____ | l | = _____ |
| m | = _____ | n | = _____ | o | = _____ | p | = _____ |

**2** The sum of the angles of any triangle equals 180°.

Using the space below, and practice paper if you wish, carry out an investigation to find out the sum of the angles of other polygons. Something to think about: *Is there a connection between the sum of the angles on a 3-, 4-, 5-, 6-sided shape?*

If you have access to a drawing program on a computer, you could use it to carry out your investigation.

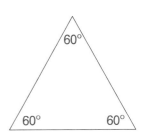

## Practice

1   Follow the instructions to make patterns.

   a   Reflect this triangle.

   b   Now, translate the triangle.

   c   Now, rotate the triangle a quarter of a turn clockwise.

2   What method of transformation has been used for the following pictures?

   a

   ![row of temple/ladder shapes]

   _____

   b

   ![row of zig-zag shapes]

   _____

   c

   ![row of windsurfer shapes]

   _____

3   Circle the sentence that best describes this transformation pattern.

   •   The shape could have been translated or reflected.

   •   The shape could have been reflected or rotated.

   •   The shape could have been translated, reflected or rotated.

   •   The shape could have been rotated or translated.

Oxford University Press

# Challenge

**1** Make a pattern by reflecting this shape horizontally and vertically.

**2** Make a pattern by translating this shape horizontally and vertically.

**3** **a** Make a pattern by rotating the following shape a quarter-turn along the top row.

   **b** Finish the pattern by reflecting the top row onto the second row.

**4** **a** Make a pattern by rotating the following shape a quarter-turn along the top row.

   **b** Finish the pattern by reflecting the top row onto the second row.

**Symmetrical faces—spot the difference**

**1** Do you have a symmetrical head? A lot of people think they do, but not many people's heads are truly symmetrical.

Look at the two pictures of the woman's head. In the first photo, her head might appear symmetrical, but compare it with the second photo.

In the second photo, the left-hand side of the woman's face has been copied, reflected and pasted onto the right-hand side.

Identify the feature(s) in the first picture that make the woman's head asymmetrical.

_____

_____

**2** Use the grid to help you to reflect the drawing of the left part of a head. Make the drawing as symmetrical as you can.

**3** Find a picture you like of something that is symmetrical or near-symmetrical. Cut it in half and draw the reflected part of the image.

Using grid lines can help you with the accuracy of your drawings.

Oxford University Press

## Practice

**1** Give the coordinate points for the:

a blue dot

_____

b purple dot

_____

c green dot

_____

d yellow dot

_____

e red dot

_____

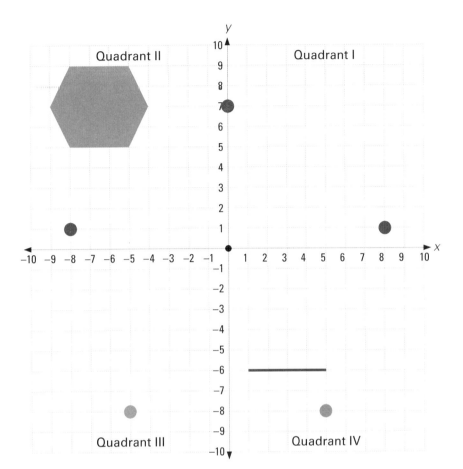

**2** Write the ordered pairs for the hexagon in question 1. Use an arrow to show the points that are being joined. The first two have been done for you:

$(-8,9) \rightarrow (-5,9) \rightarrow$ _____

**3** a Join the five coloured dots in question 1 to make a pentagon.

b Write the ordered pairs that would tell someone how to join the dots to make the pentagon.

_____

**4** a Using the red line as the base line, draw an isosceles triangle in Quadrant IV.

b Write the ordered pairs that would tell someone how to join the dots to make the triangle.

_____

**1**  **a**  Write the ordered pairs that would tell someone how to join the dots to draw the aeroplane shape in Quadrant II.

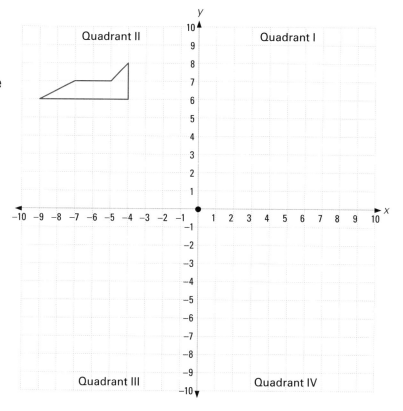

**b**  Reflect the plane into Quadrant I so that the front tip is at (9,6).

**c**  Write the ordered pairs that would tell someone how to join the dots to draw the aeroplane in Quadrant I.

**2**  Follow the ordered pairs to draw the left-hand side of a space rocket onto the grid above.

(0,10) → (−1,9) → (−1,3) → (−2,2) → (−2,−4) → (−4,−5) → (−4,−9) → (−2,−8) → (−1,−7) → (0,−7)

**3**  **a**  Reflect the drawing from question 2 into Quadrants I and IV.

**b**  Write the ordered pairs that would tell someone how to join the dots to draw the right-hand side of the rocket.

Oxford University Press

**1** On this number plane we can see that the points that the red line passes through are all in Quadrant I. Using ordered pairs, we could write:

$(0,10) \rightarrow (1,8) \rightarrow (2,6) \rightarrow (3,4)$ and so on.

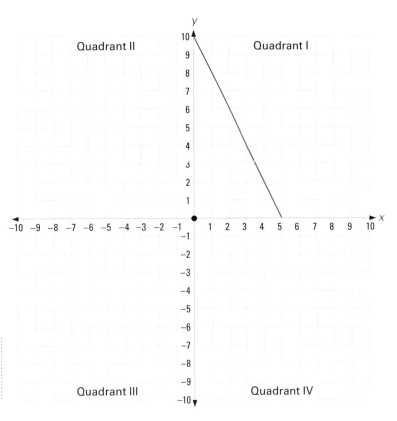

A different way is to use $x$ for the position on the $x$-axis and $y$ for the position on the $y$-axis. We can then write the coordinate points in a table of values:

| x | 0 | 1 | 2 | 3 | 4 | 5 |
|---|---|---|---|---|---|---|
| y | 10 | 8 | 6 | | | |

Complete the table of values.

**2** In order to write a rule for the table of values in question 1, we could say:

*To find the position of* y *you start with 10 and subtract double the value of* x.

This rule might sound very complicated, but we can turn it into a simple equation:

$y = 10 - (x \times 2)$.

You can test the rule by using 3 as the value of $x$: $y = 10 - 3 \times 2$.

So, remembering the order of operations, this means the position of $y$ should be 4, and this is correct.

a Using this rule, continue the table of values as far as $x = 10$

| x | 0 | 1 | 2 | 3 | 4 | 5 | 6 | 7 | 8 | 9 | 10 |
|---|---|---|---|---|---|---|---|---|---|---|----|
| y | 10 | 8 | 6 | | | | | | | | |

b Into which quadrant does the line pass? _____

c Continue the red line using the new table of values.

**3** For a different line we need a different equation.

a Using $y = -10 + (x \times 2)$, complete the table of values:

| x | 0 | 1 | 2 | 3 | 4 | 5 | 6 | 7 | 8 | 9 | 10 |
|---|---|---|---|---|---|---|---|---|---|---|----|
| y | -10 | -8 | | | | | | | | | |

b Plot the points on the number plane above and draw a line to connect them.

## Practice

**1**    Class 6G looked after the school garden. They had an open week. They made a frequency table and then a pictograph to show the number of visitors that came.

     **a**    Use the information to fill in the gaps in the frequency table and in the pictograph.

| Frequency table: The number of people who came to visit the school garden | | | | | |
|---|---|---|---|---|---|
| Day: | Monday | Tuesday | Wednesday | Thursday | Friday |
| Number: | | 14 | 21 | | |

     **b**    They gave each visitor a flower. How many flowers did they give away?

Graph to show the number of visitors to our garden in open week.

Key = 4 people

   Monday    Tuesday    Wednesday    Thursday    Friday

**2**    **a**    Transfer the data from question 1 onto a line graph. Decide on a suitable scale for the graph.

     **b**    Write a title on the graph and label the vertical axis.

     **c**    What trend about the number of visitors does the line graph show?

   Monday    Tuesday    Wednesday    Thursday    Friday

     **d**    If the garden were open for a sixth day, what would be your prediction for the number of visitors? Give a reason for your answer.

Oxford University Press

## Challenge

**1** The populations of five countries are compared on this pie graph (sometimes known as a circle graph or a sector graph).

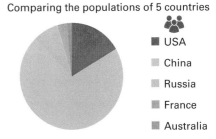

Comparing the populations of 5 countries
- USA
- China
- Russia
- France
- Australia

**a** The population of which country has a population that is almost three-quarters of the total population of the five countries? _____

**b** Write a statement of finding from the information in the graph.

_____

**2** The same data is shown on this column graph.

**a** One advantage of a pie graph is that it is an easy way of comparing data. What is an advantage of presenting the same data on a column graph?

_____

_____

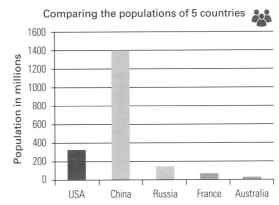

Comparing the populations of 5 countries

Population in millions (USA, China, Russia, France, Australia)

**b** From the graph we can see that the population of China is about 1400 million. Write the population in figures. _____

**c** Estimate the population of the USA. _____

**3** This graph compares the populations of pet cats in the same countries. Write a statement of finding about the pet cat population of the USA.

_____

_____

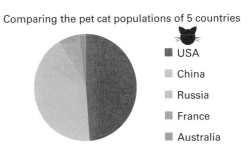

Comparing the pet cat populations of 5 countries
- USA
- China
- Russia
- France
- Australia

**4** This column graph also represents the data for the pet cat populations.

From the data we can see that the USA appears to be a nation of cat lovers.
If someone made a statement of finding that, "Around one person in four in the USA has a pet cat", how accurate would that statement be?

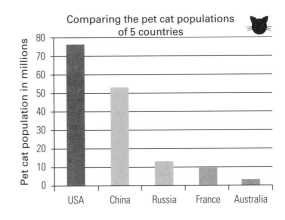

Comparing the pet cat populations of 5 countries

Pet cat population in millions (USA, China, Russia, France, Australia)

## Mastery

**1** Making a graph is a relatively simple method of representing data in an interesting way. For example, the table below shows the countries that produce the most gold. Either with pen and paper or by using a program such as Excel, this information can quickly be turned into a graph.

| The top gold-producing countries | |
|---|---|
| **Country** | **kg of gold per year** |
| China | 362 000 kg |
| Australia | 258 000 kg |
| United States | 234 000 kg |
| Russia | 199 442 kg |
| South Africa | 181 000 kg |

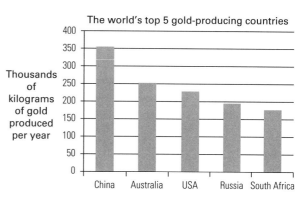

a Sometimes it is necessary to round figures to make the data easier to handle. Which gold-production figure do you think would need to be rounded and how would it be rounded?

b Notice that the figures on the vertical axis do not show the number in full—they say 400, 350 instead of 400 000, 350 000 etc. Why do you think this is?

**2** Use some of the data in the following tables to make a graph.

| Countries with the most threatened mammals | | |
|---|---|---|
| **Country** | **In 2010** | **In 2015** |
| Indonesia | 183 | 186 |
| Madagascar | 63 | 118 |
| Mexico | 99 | 101 |
| India | 93 | 96 |
| Brazil | 79 | 81 |

| Top banana-producing countries | |
|---|---|
| **Country** | **Bananas (tonnes)** |
| India | 27 575 000 t |
| China | 12 075 238 t |
| Philippines | 8 645 749 t |
| Brazil | 6 892 622 t |
| Ecuador | 5 995 527 t |

| Cities with the most skyscraper buildings | | |
|---|---|---|
| **City** | **Skyscrapers** | **Population** |
| Hong Kong, China | 1268 | 6 980 412 |
| New York City, USA | 600 | 8 274 527 |
| Tokyo, Japan | 411 | 9 262 046 |
| Chicago, USA | 295 | 4 553 009 |
| Dubai, UAE | 249 | 10 356 000 |

| Countries with the most tourists per year | |
|---|---|
| **Country** | **Number of tourists** |
| France | 83.01 million |
| USA | 66.96 million |
| China | 57.72 million |
| Spain | 56.70 million |
| Italy | 46.36 million |

Oxford University Press

# UNIT 9: TOPIC 2
## Data in the media

**Should students use smartphones in class?**

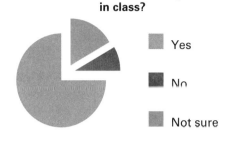

1   Students in a Year 6 class wanted to be allowed to use smartphones in class. They showed their teacher the results of a survey and said that it was clear that most people thought it was a good idea.

   a   About what percentage of people thought that smartphones should be allowed in class? _____

   b   24 people were surveyed. Four replied, "Not sure". Estimate the number who answered, "No". _____

2   The teacher asked who had been surveyed. The students answered that all the students in the class had been surveyed. Was the graph based on primary or secondary data?

   _____

3   The teacher said that they should survey the students' parents, but suggested that parents would probably think the smartphones would be used for playing games. Write a survey question that would be suitable for surveying the parents.

   _____

4   The teacher asked a few students to do some research to find out what other people thought about the idea.

   a   Circle the most suitable subject to search for online:

   •   *Should smartphones be used in classrooms?*

   •   *Should smartphones be banned in classrooms?*

   •   *Should Year 6 students be allowed to use smartphones in classrooms?*

   b   Give a reason for your choice.

   _____

5   The result of the survey of the parents was made into a graph. Estimate the percentages for *No* and *Yes*.

   _____

**Do parents think students should use smartphones in class?**

**1** The teacher of the Year 6 class on page 95 asked what survey question the students had used. It was, "Should we be able to use the calculator function on our smartphones in class?" Comment on the appropriateness of the survey question.

**2** A local newspaper heard about the graph and published it with the headline:

### SHOCKING NEWS:
Most parents want their children to use smartphones in class!

One of the Year 6 parents angrily contacted the newspaper and said that the story was completely untrue. Was it? Circle one response below. The headline was:

- based on fact and completely true
- completely untrue
- based on fact and partly true
- based on fact and possibly true.

**3** The newspaper editor decided to do some research of her own. She asked people to respond by telephoning a survey line.

Should students be allowed to play with their smartphones in class? Press 1 for *Yes* and 2 for *No*. Call NOW.

Was the editor's survey a census survey or a sample survey? _____

**4** A reporter did some research and read that, after they banned mobile phones at some schools in England, test scores for students aged 16 rose by 6.4 per cent. He published an article with the headline:

Recent survey shows that students do better when smartphones are banned!

Comment on the fairness of the article.

**5** The result of the newspaper editor's telephone poll was published the next day.

**a** 102 people answered the survey. Estimate the numbers who responded *Yes* and *No*. _____

**b** Comment on the difference between the newspaper editor's survey question (question 3) and the graph title. _____

**Should students use smartphones in class?**

■ Yes
■ No

Oxford University Press

**1** There are different opinions about when and if children should be allowed to own a smartphone.

A British newspaper reported in 2017 that billionaire Microsoft founder, Bill Gates, "… banned his children from mobile phones until they turned 14".

Others believe that smartphones are a good idea for children, because they keep students happy.

    **a** What might be another reason why students should be allowed to use a smartphone?

    **b** Why do you think Bill Gates believes that children under 14 should not have a smartphone?

**2** In June 2017, SBS TV reported that "… one third of pre-schoolers own smartphones". What is your opinion about pre-schoolers having their own smartphones?

**3** Whatever a person's viewpoint, the fact is that there has been a rise in smartphone ownership by children since 2010.

    **a** Estimate the percentage of children who owned a smartphone in 2010.

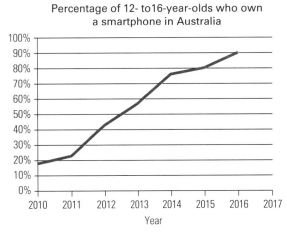

Percentage of 12- to16-year-olds who own a smartphone in Australia

    **b** The line graph shows a definite trend in smartphone ownership by children. Comment on the trend and give your opinion of the reason for it.

    **c** No figures were available for 2017. What is your estimate for the percentage?

    **d** With your teacher's permission, carry out some research to find out about smartphone ownership in your class. Decide how best to represent your data.

## Practice

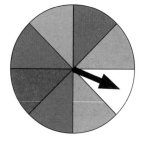

**1** a  The spinner has a one-in-how-many chance of landing on white? _____

b  What chance does the spinner have of landing on green? Answer using a percentage. _____

c  What chance does the spinner have of **not** landing on green? Answer using a decimal. _____

d  There is half as much chance of the spinner landing on yellow as on green. Describe the percentage chance for yellow. _____

e  For which colour is there a 0.375 chance? _____

**2** Describe something in your school that has the following chance of happening:

a  2% chance _____

b  50–50 chance _____

c  0% chance _____

d  0.99 of a chance _____

e  $\frac{1}{4}$ of a chance _____

**3** Which of these describes the chance of this spinner landing on red? Circle any that are appropriate.

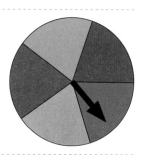

$\frac{1}{4}$   $\frac{1}{5}$   20%   25%   0.25   0.2   0.5   1 in 5   $\frac{2}{10}$

**4** A coach says the team has a very good chance of winning a game. Give a percentage value for the chance.

_____

**1**  **a**  Colour this spinner so that the following probabilities are true:

- There is a 20% chance for green.
- There is 0.3 of a chance for red.
- There is a 5% chance for purple.
- There is $\frac{1}{5}$ of a chance for yellow.
- There is 0.15 of a chance for blue.

**b**  Describe the probability of the spinner landing on the remaining white part of the spinner as a fraction, as a decimal and as a percentage. _____

---

**2**  Each jar below contains 60 jelly beans, of all colours. The number of black jelly beans in each jar is shown on the label. Write a value to show the probability of picking out a black jelly bean first time if you chose one without looking. Choose from the list.

$\frac{1}{5}$    0.25    $\frac{2}{3}$    75%    0.33̇    $\frac{7}{12}$

| A | B | C | D | E | F |
|---|---|---|---|---|---|
| 40 black jelly beans | 45 black jelly beans | 12 black jelly beans | 35 black jelly beans | 15 black jelly beans | 20 black jelly beans |
| _____ | _____ | _____ | _____ | _____ | _____ |

---

**3**  There is a mix of red and green beads in each bag of 50. Evie takes 20 from each bag without looking and records the results. Predict the number of red and green beads in each bag of 50.

| Bag | After 20 have been taken out: | | My prediction after 50 have been taken out: | |
|-----|---------|-----------|---------|-----------|
| A | Red: 16 | Green: 4 | Red: | Green: |
| B | Red: 2 | Green: 18 | Red: | Green: |
| C | Red: 6 | Green: 14 | Red: | Green: |
| D | Red: 12 | Green: 8 | Red: | Green: |

Bag A    Bag B

Bag C    Bag D

In a normal pack of playing cards there are four types, called *suits*. There are 13 cards in each suit, making a total of 52. Each suit has the same cards.

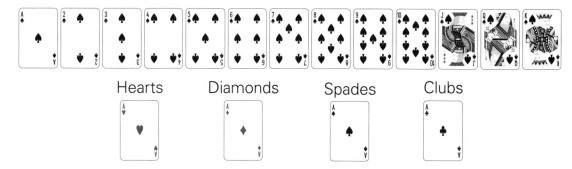

Hearts     Diamonds     Spades     Clubs

**1** For the following tasks, decide whether the values are more appropriately expressed as fractions, decimals or percentages. Imagine that the cards are turned over and shuffled around, and that somebody has to choose one without looking. What is the chance of the person guessing correctly if they predict that the card will be:

a    a red card? _____

b    a number 2? _____

c    an odd-numbered card (counting the ace as one)? _____

d    a red 4? _____

e    a diamond? _____

f    a picture card? _____

g    a picture card with a jack or a king? _____

h    a card lower than 5 (counting the ace as one)? _____

**2** There is a game called *Twenty-one* in which people try to get a total of 21 with their cards. In this game, the ace can be worth 1 or 11. The quickest way to get to 21 is to choose an ace and then a 10. The picture cards are worth 10 each, so an ace plus a 10 or a picture card makes 21.

a    What is the chance of choosing an ace on the first try? _____

b    If you choose an ace on the first try, what is the chance of choosing a card worth 10 with your next pick? Remember that there won't be 52 cards left to choose from! _____

c    Express the chance of the second card being worth 10 as a percentage (to the nearest whole number). _____

**3** Carry out some investigations by looking at the 52 cards. You could write some questions to give to others in the class, but make sure you know the answer yourself! For example, if you picked up two picture cards, what would be the probability of getting an ace to make 21 on the next try?

# UNIT 10: TOPIC 2
## Conducting chance experiments and analysing outcomes

You will need a pack of playing cards.

### Game one: Finding the right queen

**1** For this game you need the four queens from a pack of cards. Turn them face down and shuffle them around. The chance of picking out any queen is obviously 100%.

a What is the percentage chance of choosing the Queen of Diamonds? _____

b What is the decimal chance of choosing a red queen? _____

c Write the chance of **not** choosing the Queen of Spades as a fraction. _____

**2** a Still using only the four queens, choose a card at random 20 times. How many times would you expect to choose the queen from each suit? _____

b What fraction of the number of times would that be for each suit? _____

c Carry out the experiment. Record your results with tally marks in the table below.

| Queen | Tally | Tally total | Write as a fraction of 20 |
|-------|-------|-------------|---------------------------|
| Spades | | | |
| Clubs | | | |
| Hearts | | | |
| Diamonds | | | |

d How did your prediction compare with the actual results?

_____

### Game two: Kings and queens

**3** For this game you will need four queens and four kings.

a Repeat the experiment from question 2, choosing a card at random 20 times. Predict the number of times you will choose a king or a queen. _____

b Carry out the experiment. Record the results with tally marks in the table below.

| Card | Tally | Tally total | Write as a fraction of 20 |
|------|-------|-------------|---------------------------|
| A queen | | | |
| A king | | | |

c Was your prediction closer to the actual number than it was in question 2? Suggest a reason. _____

# Challenge

## Game three: Jack of Diamonds

For this game, you will need four queens and the Jack of Diamonds.

**1** Write a probability value for choosing:

a queen _____ the Jack of Diamonds _____ a black queen _____

**2** **a** Repeat the experiment from question 2 on the previous page, this time using the four queens and the Jack of Diamonds. How many times would you expect each of the following to appear?

a queen _____ a black queen _____ the Jack of Diamonds _____

**b** Repeat the experiment choosing from the five cards 20 times.

| Card | Tally | Tally total | Write as a fraction of 20 |
|---|---|---|---|
| A queen | | | |
| A black queen | | | |
| Jack of Diamonds | | | |

**c** How did your results compare to those of a classmate?

_____

**3** You need the following cards:

Hearts: 2, 3, 4 and 5     Spades: 6, 7 and 8

Clubs: 9 and 10     Diamonds: Jack of Diamonds

Write the percentage chance for choosing:

**a** a heart _____     **b** a spade _____

**c** a club _____     **d** a diamond _____

**e** a red card _____     **f** an even-numbered spade _____

**4** If you were to add the Ace of Diamonds to the 10 cards in question 3, would the chance of choosing the following be higher or lower than it is in question 3?

**a** a heart _____     **b** a spade _____

**c** a club _____     **d** a diamond _____

**e** a red card _____     **f** the Jack of Diamonds _____

**5** Write the chance for each outcome in question 4 as a fraction, a decimal (to two decimal places) and a percentage. _____

Oxford University Press

### Game four: Jumps and slides

**1** **You need:**

Two or more players, and a pencil and paper for each player. You also need a random card generator on a computer for a full pack of 52 cards.

**The game:**

The aim of the game is to get to 100.

> You could use a hundred grid for each person to keep track.

**Rules:**

Players take turns to generate a new card and jump forward by the number on the card.

An ace counts as 1 or 11. Queens and kings count as 15.

However, there are **jumps** and **slides**:

- Any card that is a multiple of 5: jump forward an extra five steps.
- Any card that is a multiple of 3: slide back three steps.
- If you land on a prime number: slide back to the previous prime number.
- Any ace: jump forward an extra 10 steps.
- Jack of Hearts, Clubs or Spades: slide back 10 steps (or to zero).
- Jack of Diamonds is the **Star Card**: jump straight to 100—you win!

If you decide to change the rules, make sure all players agree. For example, if you are up to 96, do you have to get exactly 4 to get to 100, or will any number higher than 3 allow you to win?

--------------------------------------------------------------

**2** Make up your own game using playing cards. In the game above, there was only a 1-in-52 chance of getting the Star Card. How could you increase the chances of getting a Star Card? When you have invented your game, write out the rules and then try it out to see if the rules work.

# ANSWERS

## Practice

**1**
**a** 700 000 **b** 50 000
**c** 6 000 000 **d** 10 000 000
**e** 800 **f** 200 000 000

**2**
**a** seven hundred and thirty-four thousand, eight hundred and fifteen
**b** sixty-two million, seven hundred and fifty-nine thousand, three hundred and seventy-seven
**c** six million, two hundred and nineteen thousand, seven hundred and thirty
**d** thirteen million, five hundred and eighty-five thousand, one hundred and four
**e** five million, four hundred and eighty-seven thousand, eight hundred
**f** two hundred and eighteen million, eight hundred and nineteen thousand, nine hundred and ninety-nine

**3**
**a** 413 629 = 400 000 + 10 000 + 3000 + 600 + 20 + 9
**b** 3 746 123 = 3 000 000 + 700 000 + 40 000 + 6000 + 100 + 20 + 3
**c** 52 065 350 = 50 000 000 + 2 000 000 + 60 000 + 5000 + 300 + 50
**d** 43 200 806 = 40 000 000 + 3 000 000 + 200 000 + 800 + 6
**e** 680 405 020 = 600 000 000 + 80 000 000 + 400 000 + 5000 + 20

## Challenge

**1**
**a** 10 235 789
**b** 19 875 320
**c** 123 578.9 (The zero is "non-significant".)
**d** 53 210.987

**2**
**a** 31 084 509 matches to spike abacus B.
**b** Three hundred and ten million, eighty thousand, four hundred and fifty-nine matches to spike abacus D.
**c** 310 084 509 matches to spike abacus A.
**d** Three hundred and one million, eight hundred and four thousand, five hundred and nine matches to spike abacus C.

**3**
**a** $395 000
**b** $1 300 000
**c** $910 000
**d** $1 500 000

## Mastery

**1** Students should circle the following:
3 878 224  3 949 999
3 898 100  3 918 885

**2** 23.009M, 22.909M, 22.985M

---

**3** There are 24 possible answers. Look for students who recognise that the 2 digit must be in the millions place, the 3 digit in the hundreds-of-thousands place and that, in order to round to the nearest hundred thousand, the tens-of-thousands digit must be lower than 5.
Look also for those who adopt a systematic approach to choosing numbers, e.g.

| | | | |
|---|---|---|---|
| 2 315 679 | 2 316 579 | 2 317 569 | 2 319 567 |
| 2 315 697 | 2 316 597 | 2 317 596 | 2 319 576 |
| 2 315 769 | 2 316 759 | 2 317 659 | 2 319 657 |
| 2 315 796 | 2 316 795 | 2 317 695 | 2 319 675 |
| 2 315 967 | 2 316 957 | 2 317 956 | 2 319 756 |
| 2 315 976 | 2 316 975 | 2 317 965 | 2 319 765 |

---

## Practice

**1** Student shades:
1, 4, 9, 16, 25, 36, 49, 64, 81 and 100.
**2** There is an equal number of odd and even square numbers.
**3** Students circle 5 × 5 and $5^2$.
**4** Students circle 1, 3, 6, 10, 15, 21, 28, 36, 45, 55.
**5** 1 and 36.

## Challenge

**1** 1 + 3 + 5 + 7 + 9 + 11 + 13 + 15 + 17 + 19 + 21
**2**
**a** 169
**b** 1 + 3 + 5 + 7 + 9 + 11 + 13 + 15 + 17 + 19 + 21 + 23 + 25
**3** 11
**4**

1 + 2 + 3 + 4 + 5 + 6 + 7 + 8 + 9 + 10 + 11 = 66

## Mastery

**1**
**a** 3rd TN = 3 × (3 + 1) ÷ 2 = 6
**b** 5th TN = 5 × (5 + 1) ÷ 2 = 30 ÷ 2 = 15

**2** Answers will vary depending on the level of motivation of the students and/or the teacher.

20th TN = 20 × (20 + 1) ÷ 2 = 420 ÷ 2 = 210

Students could be encouraged to use a calculator for larger triangular numbers, such as the 60th (1830) and 100th (5050) triangular numbers.

---

## Practice

**1**

| Number | Factors (numbers it can be divided by) | How many factors? | Is it prime or composite? |
|---|---|---|---|
| 21 | 1, 21, 3, 7 | 4 | composite |
| 22 | 1, 22, 2, 11 | 4 | composite |
| 23 | 1, 23 | 2 | prime |
| 24 | 1, 24, 2, 12, 3, 8, 4, 6 | 8 | composite |
| 25 | 1, 25, 5 | 3 | composite |
| 26 | 1, 26, 2, 13, | 4 | composite |
| 27 | 1, 27, 3, 9 | 4 | composite |
| 28 | 1, 28, 2, 14, 4, 7 | 6 | composite |
| 29 | 1, 29 | 2 | prime |
| 30 | 1, 3, 10, 30, 2, 15, 5, 6 | 8 | composite |

**2** The factors of 12 are 1, 12, 2, 6, 3 and 4.
**3** **a** 28 **b** 24 and 30 **4** 2
**5** **a** 25 is the square number on the chart. It has 3 factors.
**b** 4 or 9

## Challenge

**1** **a**  **b**

The prime factors of 14 are 2 and 7. So, 2 × 7 = 14

The prime factors of 77 are 7 and 11. So, 7 × 11 = 77

**c**

The prime factors of 30 are 2, 3 and 5. So, 2 × 3 × 5 = 30

Oxford University Press

**2 a**  **b**

$68 = 2 \times 2 \times 17$

$68 = 2^2 \times 17$

**c**  **d**

$56 = 7 \times 2 \times 2 \times 2$

$56 = 7 \times 2^3$

$27 = 3 \times 3 \times 3$

$27 = 3^3$

$40 = 2 \times 2 \times 2 \times 5$

$40 = 2^3 \times 5$

## Mastery

**1** Answers may vary, but students will demonstrate mastery by explaining that 37 has only two factors, 1 and itself.

**2**

| | Prime factors | Index form | Number |
|---|---|---|---|
| **a** | $2 \times 2 \times 2 \times 3$ | $2^3 \times 3$ | 24 |
| **b** | $2 \times 2 \times 2 \times 5 \times 5$ | $2^3 \times 5^2$ | 200 |
| **c** | $2 \times 2 \times 2 \times 2 \times 5$ | $2^4 \times 5$ | 80 |
| **d** | $2 \times 2 \times 3 \times 7$ | $2^2 \times 3 \times 7$ | 84 |

**3 a** 4   **b** 2   **c** 7   **d** 3

**4 a** 6   **b** 8   **c** 9

**5**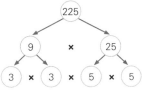

$225 = 3 \times 3 \times 5 \times 5$

$225 = 3^2 \times 5^2$

$\sqrt{225} = 3 \times 5 = 15$

Check: $15 \times 15 = 225$

## UNIT 1: Topic 4

### Practice

**1 a** 658   **b** 659   **c** 1358
   **d** 2459   **e** 4999   **f** 12 989

**2 a** 239   **b** 729   **c** 1060
   **d** 267   **e** 567   **f** 1401

**3** Teachers may prefer to ask students to compare the strategies they used with a partner or other group members. Examples of strategies that might be used include:
   **a** "I added 350 to 551 and then took away 2." Answer: 899
   **b** "I took 300 away and then added 1 back." Answer: 1173

**c** "I added 350 and 350 and then added the 700 to 8000." Answer: 8700
**d** "I took away 500 from 24 746 = 24 246. Then I took away another 20 = 24 226." Answer: 24 226

## Challenge

**1** Students could be asked to explain to each other the way that they chose to round the amounts and to justify their chosen method of rounding. Examples of the way the amounts might be rounded include:

| | Price | Rounded price | I rounded to the nearest … |
|---|---|---|---|
| **a** | $450 300 000 | $450 000 000 | million |
| **b** | $179 000 000 | $180 000 000 | ten million |
| **c** | $170 400 000 | $170 000 000 | million |
| **d** | $142 400 000 | $140 000 000 | ten million |
| **e** | $119 922 500 | $120 000 000 | million |

**2 a** $593M (Actual amount = $592 700 000)   **b** $8 600 000
   **c** $77 500   **d** $312 800 000

**3** Here are some examples of ways that rounding could be used to estimate the answers, and the exact answers:

| | Problem | Round the numbers | Estimate the answer | Calculator answer |
|---|---|---|---|---|
| **a** | 3085 + 1877 | 3000 + 2000 | 5000 | 4962 |
| **b** | 6018 – 3119 | 6000 – 3000 | 3000 | 2899 |
| **c** | 35 056 + 4023 | 35 000 + 4000 | 39 000 | 39 079 |
| **d** | 29 856 – 20 028 | 30 000 – 20 000 | 10 000 | 9828 |
| **e** | 400 247 + 49 588 | 400 000 + 50 000 | 450 000 | 449 835 |
| **f** | 398 108 – 197 434 | 400 000 – 200 000 | 200 000 | 200 674 |
| **g** | 5 498 858 – 2 889 917 | 5 500 000 – 3 000 000 | 2 500 000 | 2 608 941 |

## Mastery

**1 a** $2600
   **b** $2605

**2 a** Look for students who demonstrate an understanding of the connection between addition and subtraction. Teachers will probably wish to encourage students to use a calculator to subtract 123456789 from 987654321, and they will arrive at an answer of 864197532.
   **b** Students who have a good understanding of place value will write the answer as 864 197 532. In words this is eight hundred and sixty-four million, one hundred and ninety-seven thousand, five hundred and thirty-two.

**3** Look for students who demonstrate time-saving strategies, such as looking for addition pairs that total 1000. Possible solutions are:

| Performance | Number of people |
|---|---|
| 1 | 499 |
| 2 | 501 |
| 3 | 498 |
| 4 | 502 |
| 5 | 497 |
| 6 | 503 |
| 7 | 496 |
| 8 | 504 |
| Total | 4000 |

## UNIT 1: Topic 5

### Practice

**1 a** 223   **b** 456
   **c** 789   **d** 1234
   **e** 12 345   **f** 123 456
   **g** 1 234 567   **h** 12 345 678
   **i** 123 456 789

### Challenge

**1 a**

| | Country | Length of paved roads (km) | Length of unpaved roads (km) | Total length of roads (km) |
|---|---|---|---|---|
| 1 | USA | 4 165 110 | 2 265 256 | 6 430 366 |
| 2 | India | 1 603 705 | 1 779 639 | 3 383 344 |
| 3 | China | 1 515 797 | 354 864 | 1 870 661 |
| 4 | Japan | 925 000 | 258 000 | 1 183 000 |
| 5 | Russia | 738 000 | 133 000 | 871 000 |
| 6 | Spain | 659 629 | 6663 | 666 292 |
| 7 | Canada | 415 600 | 626 700 | 1 042 300 |
| 8 | Australia | 336 962 | 473 679 | 810 641 |
| 9 | Brazil | 96 353 | 1 655 515 | 1 751 868 |

**b** USA, India, China, Brazil, Japan, Canada, Russia, Australia, Spain

# ANSWERS

## Mastery

In this section look for students who use their knowledge of place value and who see the connection between addition and subtraction as part of their problem-solving strategies.

1. Students could subtract any 5-digit number from 123 321, finding two addends that would give the correct total. For example, 123 321 – 52 897 = 70 424, so the two addends would be 52 897 + 70 424 = 123 321. A similar process could be used for the other two sums. Those looking for a short cut to find the other two solutions could subtract one from one addend and add one to the other to solve the problem (i.e. 52 896 + 70 425 and 52 895 + 70 426).

2. This problem can be solved in many ways. Perhaps the easiest would be to use a knowledge of expanded numbers, renaming 123 456 789 as:

   123 000 000 + 400 000 + 50 000 + 6789.

3. The question does not specify the number of addends, but students will probably not choose to have a high number. They will probably use a process of trial and error to solve the problem. Look for students who see that, since the ones column should total 6, one of the addends will end in 5 and another in 1. One possible solution is 10 755 + 1171 + 120.

4. As with question 1, students who have a good grasp of place value will probably use their knowledge of expanded numbers to solve the problem. For example, they could start with the answer and subtract 21 000 000 from it, leaving 966 654 321. 21 000 000 could then be expanded into 19 000 000 + 1 900 000 + 100 000. The resulting addition would be 966 654 321 + 19 000 000 + 1 900 000 + 100 000 = 987 654 321.

## UNIT 1: Topic 6

### Practice

1.
   a. 121 212
   b. 232 323
   c. 343 434
   d. 454 545
   e. 565 656
   f. 676 767
   g. 787 878
   h. 898 989
   i. 989 898
   j. 878 787

2. Students who see the connection between subtraction and addition will solve the problem by adding 767 676 and 343 965 to find that the number on the top line is 1 111 641.

### Challenge

1.
   a. 321 321
   b. 432 432
   c. 43 214 321
   d. 54 325 432
   e. 65 436 543
   f. 76 547 654
   g. 87 658 765
   h. 98 769 876
   i. 123 454 321
   j. 543 212 345

2. There are many different solutions that are possible. Students will probably use trial and error, beginning by writing a top line with six zeros, e.g. 300 400 700. They can then subtract 123 456 789 from the number to work out the number that needs to appear on the second line of the subtraction.

In this example the subtraction would become 300 400 700 – 176 943 911 = 123 456 789.

### Mastery

1. 11 933 tonnes
2.
   a. Colombia
   b. 60 277 t
   c. 156 675 t
   d. 1 821 754 t
   e. 2 300 889 t – 2 194 172 t = 106 718 t
   f. 4 495 060 t – 3 545 424 t = 949 636 t

## UNIT 1: Topic 7

### Practice

1.

|   |      | ×10     | ×100    | ×1000    |
|---|------|---------|---------|----------|
| a | 43   | 430     | 4300    | 43 000   |
| b | 87   | 870     | 8700    | 87 000   |
| c | 235  | 2350    | 23 500  | 235 000  |
| d | $7.75 | $77.50 | $775.00 | $7750.00 |
| e | 250  | 2500    | 25 000  | 250 000  |
| f | 36   | 360     | 3600    | 36 000   |
| g | $0.25 | $2.50  | $25.00  | $250.00  |
| h | $10.15 | $101.50 | $1015.00 | $10 150.00 |

2.

|   |       | ÷10   | Write the multiplication-fact partner |
|---|-------|-------|---------------------------------------|
| a | 350   | 35    | 35 × 10 = 350                         |
| b | 3600  | 360   | 360 × 10 = 3600                       |
| c | 5000  | 500   | 500 × 10 = 5000                       |
| d | $12.50 | $1.25 | $1.25 × 10 = $12.50                  |
| e | 73    | 7.3   | 7.3 × 10 = 73                         |

3.

|   |        | ÷ 100  | Write the multiplication-fact partner |
|---|--------|--------|---------------------------------------|
| a | 800    | 8      | 8 × 100 = 800                         |
| b | $255   | $2.55  | $2.55 × 100 = $255                    |
| c | 7000   | 70     | 70 × 100 = 7000                       |
| d | 15 000 | 150    | 150 × 100 = 15 000                    |
| e | 7250   | 72.5(0) | 72.5 × 100 = 7250                    |

### Challenge

1.

|   | × 10 | × 20 | × 40 | × 80 |
|---|------|------|------|------|
| a | 11   | 110  | 220  | 440  | 880  |
| b | 14   | 140  | 280  | 560  | 1120 |
| c | 21   | 210  | 420  | 840  | 1680 |
| d | 35   | 350  | 700  | 1400 | 2800 |
| e | 60   | 600  | 1200 | 2400 | 4800 |

2.

|   |        | ÷ 10 | ÷ 20 | ÷ 40 | ÷ 80 |
|---|--------|------|------|------|------|
| a | 800    | 80   | 40   | 20   | 10   |
| b | 4000   | 400  | 200  | 100  | 50   |
| c | 720    | 72   | 36   | 18   | 9    |
| d | 12 000 | 1200 | 600  | 300  | 150  |
| e | 4800   | 480  | 240  | 120  | 60   |

3.
   a. 500
   b. 960
   c. 750
   d. 560
   e. 750
   f. 2500
   g. 1500
   h. 1680
   i. 380
   j. $65
   k. $55
   l. 112
   m. 90
   n. 360

4. Answers may vary, e.g. "I multiplied 22 × 10 = 220. Then I doubled 22 and doubled it again to get 88. I added 220 and 88 = 308. So, 22 × 14 = 308."

### Mastery

1. 90 × 70 = 6300
2.
   a. Answers may vary. Look for students who see that, for numbers that are multiples of 10, the number of zeros in the answers is the same as the total number of zeros in the two numbers being multiplied.
   b. The investigation should result in proving the above to be correct. Some students may reach this conclusion after a couple of tries whereas others may wish to try many multiplications to test the theory.
3.
   a. The sentence that would not give the correct answer is, "You could halve 648 and move the digits one place bigger".
   b. Teachers may wish to give students the opportunity to share their choice of strategy with others so that students can see that what works best for one is not necessarily the choice of others.
4. This could form part of a cooperative group activity.

   5 for $107.50 = $21.50 per seat.
   32 for $624 = $19.50 per seat.

   Students will probably conclude that, even though only 30 seats are required, the school would save $21 by purchasing 32 seats from the second supplier.

## UNIT 1: Topic 8

### Practice

1.
   a. 852
   b. 957
   c. 680
   d. 738
   e. 6956
   f. 7710
   g. 9380
   h. 11 459
   i. 13 968
   j. 19 215
   k. 32 040
   l. 56 781
   m. 14 560
2. $18 468
3. 6 × $349 845 = $2 099 070 plus $337 845 = $2 436 915 in total

### Challenge

1.
   a. 520
   b. 680
   c. 810
   d. 2120
   e. 1620
   f. 2200
   g. 5190
   h. 15 180
   i. 48 580
   j. 81 400
   k. 155 700
   l. 232 470
2.
   a. 600
   b. 910
   c. 864
   d. 3792
   e. 20 020
   f. 14 256

### Mastery

1. Look for students who choose to make the calculations simpler by converting the masses to tonnes (e.g. the northern right whale has a mass of 77.7 t).
   a. The bowhead whale is almost twice the weight of the sperm whale. Bowhead whale = 86 000 kg; sperm whale = 43 700 kg (43 700 kg + 43 700 kg = 87 400kg).

Oxford University Press

**b** 1 787 100 kg or 1787.1 t

**c** 502.5 m

**d** 26 fin whales = 1 648 400 kg (1648.4 t); 12 blue whales = 1 644 000 kg (1644.0 t). Therefore, 26 fin whales have a greater mass than 12 blue whales.

**e** 15 × 77.7 t = 1165.5 t (1 165 500 kg) 15 × 43.7 t = 655.5 t (655 500 kg) The difference is 510 t (510 000 kg).

**2** Teachers might encourage students to write multiplications involving more than two digits or ask them to use a calculator. By writing a long multiplication for 111 × 111, for example, we can observe the way the pattern is made:

|   |   |   |   | 1 | 1 | 1 |
|---|---|---|---|---|---|---|
|   |   | × |   | 1 | 1 | 1 |
|   |   |   |   | 1 | 1 | 1 |
|   |   |   | 1 | 1 | 1 | 0 |
| + | 1 | 1 | 1 | 0 | 0 |   |
|   | 1 | 2 | 3 | 2 | 1 |   |

The pattern continues for 1111 × 1111 (=1 234 321), but calculators may not be available that will handle very large numbers. The pattern eventually breaks down after 111 111 111². However, students will probably have had enough by that stage!

Other patterns can be found by multiplying by 11, 111 and by 1111 instead of squaring the numbers. For example, 11 × 11 = 121, 111 × 11 = 1221, 1111 × 11 = 12 221.

## UNIT 1: Topic 9

### Practice

**1 a** 3 r1 or $3\frac{1}{4}$  **b** 7 r3 or $7\frac{3}{4}$

**c** 4 r1 or $4\frac{1}{8}$  **d** 7 r1 or $7\frac{1}{9}$

**e** 9 r3 or $9\frac{3}{8}$  **f** 8 r3 or $8\frac{3}{7}$

**g** 8 r2 or $8\frac{2}{9}$  **h** 9 r5 or $9\frac{5}{6}$

**2 a** 141 r3 and $141\frac{3}{4}$

**b** 74 r2 and $74\frac{2}{5}$

**c** 34 r2 and $34\frac{2}{6}$ ($34\frac{1}{3}$)

**d** 106 r3 and $106\frac{3}{7}$

**e** 321 r2 and $321\frac{2}{8}$ ($321\frac{1}{4}$)

**f** 194 r8 and $194\frac{8}{9}$

**g** 1034 r3 and $1034\frac{3}{4}$

**h** 472 r5 and $472\frac{5}{7}$

**3 a** 257 r1  **b** 614 r2

**c** 379 r3  **d** 647 r3

**e** 2011 r2  **f** 1007 r6

### Challenge

**1 a** 178.25  **b** 163.6

**c** 453.5  **d** 124.5

**e** 637.25  **f** 1171.25

**g** 5730.5  **h** 14 201.6

**i** 3478.75

---

**2 a** 634.75  **b** 2484.33

**c** 1411.17  **d** 7280.75

**e** 10 149.67  **f** 11 778.86

**g** 11 279.5  **h** 6041.22

**i** 3626.43

**3** Look for students who show an understanding of real-world numbers by identifying that, although 23 455 ÷ 8 = 2931.875, the actual number of boxes (of eight pencils) is 2931 (with a remainder of seven pencils).

### Mastery

**1 a** The total of the six scores is 317, making the mean average score 52.83.

**b** The new average score is 317 divided by 7 = 45.29.

**c** He must score at least 106 in the eighth innings to bring his total to 423. This would raise his average to 52.88. Students may choose to multiply 53 × 8 = 424, raising his new average to 53.

**2** Students will need to calculate the week's total as 31 × 7 = 217. The three temperatures shown total 94.5°C. If this is subtracted from 217, it means that the other four days must total 122.5. Look for students who choose realistic temperatures, such as 30.5°C, 30°C, 31°C and 31°C.

**3 a**

| Country | Amount of potatoes per person per week | Amount of potatoes per person per day |
|---|---|---|
| Belarus | 3420.41 g | 488.63 g |
| Poland | 2616.11 g | 373.73 g |
| Russia | 2513.91 g | 359.13 g |
| Ireland | 2436.07 g | 348.01 g |
| Australia | 1015.24 g | 145.03 g |

**b** It is likely that students will not all arrive at the same answer, as this will depend on the search engine. The date (and reliability) of the data could also affect the answers. The following shows a possible response:

|   | Country | Potato production |
|---|---|---|
| 1 | China | 95 941 500 t |
| 2 | India | 45 343 600 t |
| 3 | Russia | 30 199 126 t |
| 4 | Ukraine | 22 258 600 t |
| 5 | United States | 19 843 919 t |

## UNIT 1: Topic 10

### Practice

**1 a** −4 + 5 = 1  **b** 3 − 5 = −2

**c** 6 − 8 = −2  **d** 0 − 7 = −7

**e** −6 + 9 = 3  **f** −2 + 7 = 5

**g** 1 − 7 = −6  **h** 2 + 5 = 7

**i** 0 + 6 = 6  **j** 7 − 11 = −4

**k** −7 + 11 = 4  **l** 0 + 8 = 8

**2** If students are asked to check with a calculator and it is the kind that does not allow them to enter a negative number, teachers could use this as a discussion point

---

as to why their human brain is more reliable (as in Q 2b) than the electronic "brain" (or the way in which the calculator has been programmed).

**a** −5  **b** 1  **c** −25

**d** −70  **e** 25  **f** −3

### Challenge

**1 a** −45, −40, −35, −30, −25, −20, −15, −10, −5, 0, − 5, 10

**b** −24, −21, −18, −15, −12, −9, −6, −3, 0, 3, 6, 9

**c** −36, −27, −18, −9, 0, 9, 18, 27, 36, 45, 54, 63

**d** −200, −175, −150, −125, −100, −75, −50, −25, 0, 25, 50, 75

**e** −36, −30, −24, −18, −12, −6, 0, 6, 12, 18, 24, 30

**2 a** Queensland: −11°C (Stanthorpe) and 49.5°C (Birdsville)

**b** −23°C + 8°C = −15°C

**c** 55.8°

**d** NSW: −23°C (Charlotte Pass) and 50°C (Wilcannia). The difference is 73°C.

### Mastery

**1 a**

| Date | Credit $ (Money going in) | Debit $ (Money going out) | Balance $ |
|---|---|---|---|
| 1 June | 500.00 |  | 500.00 |
| 3 June |  | 125.50 | 374.50 |
| 5 June |  | 69.00 | 305.50 |
| 10 June |  | 420.10 | − 114.60 |
| 16 June |  | 81.75 | − 196.35 |
| 21 June | 325.00 |  | 128.65 |
| 30 June |  | − 348.95 | − 220.30 |

**b** $−220.30 + $500.00 = $279.70

**2 a**

| Throw | + or − | Number on dice | Player A moves | Player B moves |
|---|---|---|---|---|
| 1 | negative | 2 | −2 | +2 |
| 2 | positive | 3 | +3 | −3 |
| 3 | negative | 4 | −4 | +4 |
| 4 | negative | 1 | −1 | +1 |
| 5 | positive | 6 | +6 | −6 |

# ANSWERS

**b**

Player A

Player B

**c** Player A wins by 4.

## UNIT 2: Topic 1

### Practice

**1** $\frac{2}{6}, \frac{3}{9}, \frac{4}{12}$

**2** Teachers could ask students to justify other answers.

   **a** $\frac{2}{5}$      **b** $\frac{2}{3}$      **c** $\frac{1}{2}$

   **d** $\frac{1}{3}$      **e** $\frac{1}{5}$      **f** $\frac{3}{4}$

**3** **a** $\frac{1}{3}$      **b** $\frac{4}{6}$      **c** $\frac{8}{12}$

   **d** $\frac{2}{3}$      **e** $\frac{1}{6}$      **f** $\frac{3}{4}$

### Challenge

**1** Teacher to check shading and to decide how accurate the individual students need to be when dividing the shapes.

**2** Answers other than those shown below are possible. Teachers could ask students to justify other responses. Likely correct answers are:

   **a** $\frac{2}{12}, \frac{1}{6}$

   **b** $\frac{3}{4}$ and any equivalent faction, such as $\frac{6}{8}$

   **c** $\frac{6}{12}$ of the line is between the two shapes so students could list any fraction that is equivalent to $\frac{1}{2}$.

   **d** Students will, hopefully, recognise that $\frac{10}{12}$ is an equivalent fraction to $\frac{20}{24}$ and will draw a star above the $\frac{10}{12}$ mark.

   **e** Any fractions that are equivalent to $\frac{20}{24}$ are acceptable. In addition to $\frac{10}{12}$, students will probably identify $\frac{5}{6}$ as an equivalent fraction.

**3** **a** $\frac{2}{4} = \frac{1}{2}$. Numerator and denominator ÷ 2

   **b** $\frac{3}{12} = \frac{1}{4}$. Numerator and denominator ÷ 3

   **c** $\frac{2}{10} = \frac{1}{5}$. Numerator and denominator ÷ 2

   **d** $\frac{8}{12} = \frac{2}{3}$. Numerator and denominator ÷ 4

   **e** $\frac{5}{6} = \frac{10}{12}$. Numerator and denominator × 2

   **f** $\frac{6}{8} = \frac{3}{4}$. Numerator and denominator ÷ 2

   **g** $\frac{4}{5} = \frac{8}{10}$. Numerator and denominator × 2

   **h** $\frac{6}{9} = \frac{2}{3}$. Numerator and denominator ÷ 3

   **i** $\frac{3}{5} = \frac{6}{10}$. Numerator and denominator × 2

   **j** $\frac{15}{20} = \frac{3}{4}$. Numerator and denominator ÷ 5

### Mastery

**1** **a** $\frac{1}{5}$      **b** $\frac{1}{4}$      **c** $\frac{3}{4}$

   **d** $\frac{1}{2}$      **e** $\frac{1}{4}$      **f** $\frac{9}{10}$

**2** **a** There are 8 of the 32 slices left. The lowest equivalent form is $\frac{1}{4}$.

   **b** 24 of the 32 slices have been eaten. This could be written as $\frac{12}{16}, \frac{6}{8}$ or $\frac{3}{4}$.

---

**3** **a** Teacher to check explanation. Reducing $\frac{800}{1200}$ to its lowest equivalent form will show that two-thirds of the bricks were broken.

   **b** The refund should be $\frac{2}{3}$ of $408 = $272 ($\frac{1}{3}$ = $136, so $\frac{2}{3}$ = $272.)

   **c** If three-quarters of the bricks had been broken, there would have been 900 broken and 300 unbroken bricks.

**4** The easiest solution is to draw a rectangle. Students who show an excellent understanding of equivalent fractions will divide the rectangle into 24 ths. $\frac{1}{4}$ of the cake is $\frac{6}{24}$. This leaves $\frac{18}{24}$ to be shared between the six people. Each person will get $\frac{3}{24}$ of the cake. Reduced to its lowest equivalent form, this is $\frac{1}{8}$ of the cake each.

## UNIT 2: Topic 2

### Practice

**1** **a** $\frac{7}{10}$       **b** $\frac{7}{8}$

   **c** $\frac{3}{5}$       **d** $\frac{6}{7}$

   **e** $\frac{9}{10}$       **f** $\frac{4}{5}$

   **g** $\frac{5}{8}$       **h** $\frac{10}{10}$ or 1 (whole)

   **i** $\frac{8}{8}$ or 1 (whole)

**2** Teachers could encourage students to reduce the answers to their lowest equivalent form where appropriate.

   **a** $\frac{2}{8}$ (or $\frac{1}{4}$)  **b** $\frac{4}{10}$ (or $\frac{2}{5}$)  **c** $\frac{4}{12}$ (or $\frac{1}{3}$)

   **d** $\frac{5}{8}$      **e** $\frac{4}{12}$    **f** $\frac{2}{12}$ (or $\frac{1}{6}$)

   **g** 0       **h** $\frac{13}{20}$    **i** $\frac{10}{15}$ (or $\frac{2}{3}$)

**3** Teachers could encourage students to reduce the answers to their lowest equivalent form where appropriate.

   **a** $\frac{3}{8}$      **b** $\frac{5}{6}$      **c** $\frac{5}{10}$ (or $\frac{1}{2}$)

   **d** $\frac{7}{8}$      **e** $\frac{7}{9}$      **f** $\frac{5}{12}$

**4** Teachers could encourage students to reduce the answers to their lowest equivalent form where appropriate.

   **a** $\frac{5}{10}$ (or $\frac{1}{2}$)  **b** $\frac{2}{10}$ (or $\frac{1}{5}$)  **c** $\frac{1}{12}$

   **d** $\frac{7}{10}$      **e** $\frac{8}{12}$ (or $\frac{2}{3}$)  **f** $\frac{1}{10}$

**5** Students should show that $\frac{1}{2} + \frac{3}{8} = \frac{4}{8} + \frac{3}{8} = \frac{7}{8}$

### Challenge

Teachers could encourage students to reduce fractions to their lowest equivalent form where appropriate for the activities on this page.

---

**1** **a** $\frac{7}{10} + \frac{5}{10} = \frac{12}{10} = 1\frac{2}{10}$  **b** $\frac{5}{12} + \frac{11}{12} = \frac{16}{12} = 1\frac{4}{12}$

   **c** $\frac{8}{9} + \frac{2}{9} = \frac{10}{9} = 1\frac{1}{9}$  **d** $\frac{3}{5} + \frac{3}{5} = \frac{6}{5} = 1\frac{1}{5}$

   **e** $\frac{7}{9} + \frac{7}{9} = \frac{14}{9} = 1\frac{5}{9}$  **f** $\frac{5}{8} + \frac{5}{8} = \frac{10}{8} = 1\frac{2}{8}$

   **g** $\frac{6}{10} + \frac{7}{10} = \frac{13}{10} = 1\frac{3}{10}$

   **h** $\frac{1}{3} + \frac{2}{3} = \frac{3}{3} = 1$ (whole)

**2** **a** $1\frac{3}{8} - \frac{1}{8} = 1\frac{2}{8}$  **b** $1\frac{4}{9} - \frac{2}{9} = 1\frac{2}{9}$

   **c** $1\frac{7}{8} - \frac{4}{8} = 1\frac{3}{8}$  **d** $2\frac{3}{4} - 1\frac{1}{4} = 1\frac{2}{4}$

   **e** $1\frac{5}{8} - \frac{5}{8} = 1$ (whole)  **f** $3\frac{8}{10} - \frac{9}{10} = 2\frac{9}{10}$

   **g** $3\frac{2}{9} - \frac{5}{9} = 2\frac{6}{9}$  **h** $3\frac{1}{10} - 1\frac{8}{10} = 1\frac{3}{10}$

**3** **a** $\frac{3}{8} + \frac{6}{8} = \frac{9}{8} = 1\frac{1}{8}$  **b** $\frac{4}{10} + \frac{9}{10} = \frac{13}{10} = 1\frac{3}{10}$

   **c** $1\frac{6}{8} + \frac{5}{8} = 2\frac{3}{8}$  **d** $2\frac{4}{6} + \frac{5}{6} = 2\frac{9}{6} = 3\frac{3}{6}$

   **e** $3\frac{7}{8} + \frac{6}{8} = 4\frac{5}{8}$  **f** $2\frac{5}{6} + \frac{2}{6} = 2\frac{7}{6} = 3\frac{1}{6}$

**4** **a** $1\frac{1}{8} - \frac{2}{8} = \frac{7}{8}$  **b** $3\frac{7}{10} - \frac{5}{10} = 3\frac{2}{10}$

   **c** $2\frac{3}{12} - \frac{6}{12} = 1\frac{9}{12}$  **d** $1\frac{2}{6} - \frac{5}{6} = \frac{3}{6}$

   **e** $3\frac{1}{4} - 2\frac{2}{4} = \frac{3}{4}$  **f** $1\frac{5}{12} - \frac{8}{12} = \frac{9}{12}$

### Mastery

**1** Students could draw a diagram to show their reasoning. Alternatively, they could use their knowledge of equivalent fractions to show that $\frac{3}{5} = \frac{12}{20}$ and is not, therefore, equivalent to $\frac{15}{20}$.

**2**

| Fraction: $\frac{12}{30}$ | Factors of the numerator: 1, 2, 3, 4, <u>6</u>, 12 |
|---|---|
| | Factors of the denominator: 1, 2, 3, 5, <u>6</u>, 10, 30 |

The simplest form of $\frac{12}{30}$ is $\frac{2}{5}$.

**3** **a**

| Fraction: $\frac{18}{48}$ | Factors of the numerator: 1, 2, 3, <u>6</u>, 9, 18 |
|---|---|
| | Factors of the denominator: 1, 2, 3, <u>6</u>, 8, 12, 24, 48 |

The simplest form of $\frac{18}{48}$ is $\frac{3}{8}$.

   **b**

| Fraction: $\frac{25}{45}$ | Factors of the numerator: 1, <u>5</u>, 25 |
|---|---|
| | Factors of the denominator: 1, <u>5</u>, 9, 45 |

The simplest form of $\frac{25}{45}$ is $\frac{5}{9}$.

   **c**

| Fraction: $\frac{28}{44}$ | Factors of the numerator: 1, 2, <u>4</u>, 7, 14, 28 |
|---|---|
| | Factors of the denominator: 1, 2, <u>4</u>, 11, 22, 44 |

The simplest form of $\frac{28}{44}$ is $\frac{7}{11}$.

## UNIT 2: Topic 3

### Practice

**1** Student should shade:

   **a** any 2 squares    **b** any 25 squares

   **c** any 70 squares    **d** any 45 squares

**2** **a** 0.03 < 0.3: True

   **b** $\frac{1}{1000}$ = 0.001: True

   **c** $\frac{99}{1000}$ = 0.99: False

   **d** 0.03 > $\frac{3}{1000}$: True

   **e** $\frac{735}{1000}$ < 0.735: False

   **f** $\frac{3}{4}$ > 0.075: True

   **g** 1.35 = $1\frac{35}{100}$: True

   **h** 4.350 > 4.35: False

3 Student colours 20 squares blue, 10 squares green, 5 squares yellow, 25 squares red, 39 squares purple = 99 squares in total. The amount left unshaded is $\frac{1}{100}$ or 0.01.

4 0.137, 0.31, 0.37, 0.371, 1.037, 1.37

## Challenge

1 Students could be asked to simplify fractions where appropriate. Teachers may wish to discuss the use of the non-significant zero (see question 1f).

| | Fraction | Decimal |
|---|---|---|
| a | $\frac{9}{1000}$ | 0.009 |
| b | $\frac{4}{100}$ | 0.04 |
| c | $\frac{1}{1000}$ | 0.001 |
| d | $\frac{16}{1000}$ | 0.016 |
| e | $\frac{1}{4}$ or $\frac{25}{100}$ | 0.25 |
| f | $\frac{750}{1000}$ | 0.75 (0) |
| g | $\frac{99}{1000}$ | 0.099 |
| h | $\frac{1}{2}$ | 0.5 |

2 Teachers may wish to discuss with students the issue of reducing fractions to their lowest forms when dealing with decimals and fractions. For example, 0.8 is more commonly known as $\frac{8}{10}$ instead of $\frac{4}{5}$, whereas $\frac{25}{100}$ is commonly converted to $\frac{1}{4}$.

| | Improper fraction | Mixed number | Decimal |
|---|---|---|---|
| a | $\frac{5}{4}$ | $1\frac{1}{4}$ | 1.25 |
| b | $\frac{24}{10}$ | $2\frac{4}{10}$ | 2.4 |
| c | $\frac{175}{100}$ | $1\frac{75}{100}$ or $1\frac{3}{4}$ | 1.75 |
| d | $\frac{250}{100}$ | $2\frac{1}{2}$ or $2\frac{50}{100}$ | 2.5 |
| e | $\frac{475}{100}$ | $4\frac{3}{4}$ or $4\frac{75}{100}$ | 4.75 |
| f | $\frac{2750}{1000}$ | $2\frac{3}{4}$ or $2\frac{750}{1000}$ | 2.75 |

3 a 0.4    b 0.8
   c 0.25    d 0.75
   e 1.4    f 1.25

## Mastery

1 a 0.625   b 0.714   c 0.143
   d 0.875   e 0.583   f 0.273

2 a $0.\dot{6}$   b $0.1\dot{6}$   c $0.1\dot{8}$
   d $0.8\dot{3}$   e $0.6\dot{3}$   f $0.\dot{4}$

3 Students with an excellent understanding should not need to do a great deal of trial and error. They will, hopefully, have discovered that fractions such as fifths and tenths will have no recurring decimals. The easiest solution is to convert thirds, sixths or ninths to decimals.

4 Students may be surprised to learn that many countries use a comma as a separator between whole numbers and decimals. France, Germany and, in fact, most European countries, plus nearly all South American countries, adopt this practice. The UK, USA, Australia and much of Asia use the dot. The internationalisation of learning and the associated need to eliminate confusion is, perhaps, the primary reason we in Australia now use spaces to separate sets of three whole-number digits in the modern counting system.

---

## UNIT 2: Topic 4

### Practice

1 a 4.22   b 71.7   c 56.165
   d 48.1   e 1.77   f 375.29
   g 0.06   h 6.57   i 135.79

2 a 31.8   b 17.65   c 48.58
   d 8.81   e 8.63   f 8.6

3 a 28.86 m    b 20 kg

### Challenge

1 9.58 seconds

2 525 mL or 0.525 L

3 The total is 3.62 m so there is a piece that is 28 cm (or 0.28 m) left over.

4 11 servings can be taken (totalling 1485 g, or 1.485 kg).

5 475 g / 0.475 kg

### Mastery

1 Students will probably use a process of trial and error to solve the problem. However, look for students who recognise that the last addend must have a 4 in the ones column. The digits in the two blank tenths-columns need to total 7 as there is a 2 in the final tenths-column, and there is a tenth to be traded once the hundredths are added together. Once this step has been reached the gaps can be filled in a logical sequence:

0.1 + 5.63 + 4.27 = 10
0.2 + 5.53 + 4.27 = 10
0.3 + 5.43 + 4.27 = 10
0.4 + 5.33 + 4.27 = 10
0.5 + 5.23 + 4.27 = 10
0.6 + 5.13 + 4.27 = 10
0.7 + 5.03 + 4.27 = 10

2 Students could be encouraged to solve this problem as part of a cooperative group challenge. Hopefully they will realise that smaller digits will be needed for the whole numbers at positions a, b and c so that the sum will not be too large. (The digit at the d position can be larger.) Look for students who use mental strategies to eliminate some potential solutions before committing pencil to paper. One solution is: a = 2, b = 0, c = 4, d = 8. The addition becomes 2.0 + 4.8 + 2.4 + 0.8 = 10. Some other solutions include:

a = 1   b = 6   c = 0   d = 7
a = 2   b = 3   c = 1   d = 8
a = 2   b = 1   c = 3   d = 8
a = 1   b = 4   c = 2   d = 7

---

## UNIT 2: Topic 5

### Practice

1 a 12.3   b 2.34   c 34.5
   d 12.34   e 2.345   f 3.456
   g 456.78   h 87.654   i 98.765

---

2 a 12.1   b 2.31   c 34.3
   d 4.54   e 5.65   f 6.76
   g 7.87   h 8.98   i 9.109
   j 12.321   k 234.32   l 34.543

3 $4734 ÷ 5 = $946.80. This is $53.20 less than $1000.

### Challenge

1 a 8 484.84    b 16 262.61
   c 309.0906    d 309.0906
   e 22 727.25    f 5 090.904
   g 111.0111    h 333.0333
   i 222.0222

2 a $13.33 = $13.35
   b $18.50
   c $21.665 = $21.65
   d $12.49875 = $12.50
   e 2.66 recurring = $2.65
   f $2.85
   g $9.598 = $9.60
   h $33.325 = $33.33 or $33.35 (Possible discussion point as to how to round the number.)

### Mastery

1 Five balls will cost $29.35 × 5 = $146.75, so he can afford to buy four balls ($29.35 × 4 = $117.40).

2

| 3 | 0.1666 | 2 |
|---|---|---|
| 0.6666 | 1 | 1.5 |
| 0.5 | 6 | 0.3333 |

---

## UNIT 2: Topic 6

### Practice

Teachers to decide what is acceptable if students use non-significant zeros in their answers.

1 a 47   b 470   c 4700
2 a 93   b 930   c 9300
3 a 23.9   b 239   c 2390
4 a 30.01   b 300.1   c 3001
5 a 23.42   b 2.342   c 0.2342
6 a 42.3   b 4.23   c 0.423
7 a 723.48   b 72.348   c 7.2348
8 a 26.05   b 2.605   c 0.2605
9 a $235   b 3 tonnes   c 10.5 m
10 a $1.75   b 3.4 tonnes   c 5.242 km

# ANSWERS

## Challenge

**1**

| | × 30 | First × 10 | then × 3 | Multiplication fact |
|---|---|---|---|---|
| **a** | 1.2 | 12 | 36 | 1.2 × 30 = 36 |
| **b** | 2.3 | 23 | 69 | 2.3 × 30 = 69 |
| **c** | 3.1 | 31 | 93 | 3.1 × 30 = 93 |
| **d** | 1.5 | 15 | 45 | 1.5 × 30 = 45 |
| **e** | 0.32 | 3.2 | 9.6 | 0.32 × 30 = 9.6 |

**2**

| | × 30 | First × 3 | then × 10 | Multiplication fact |
|---|---|---|---|---|
| **a** | 1.3 | 3.9 | 39 | 1.3 × 30 = 39 |
| **b** | 2.2 | 6.6 | 66 | 2.2 × 30 = 66 |
| **c** | 3.2 | 9.6 | 96 | 3.2 × 30 = 96 |
| **d** | 5.1 | 15.3 | 153 | 5.1 × 30 = 153 |
| **e** | 0.33 | 0.99 | 9.9 | 0.33 × 30 = 9.9 |

**3** **a** 68　　**b** 100　　**c** 175
　　**d** 102　　**e** 48　　**f** 250
　　**g** 75　　**h** 144　　**i** 51
　　**j** $60(.00)　　**k** $67.50　　**l** 84 L
　　**m** 249 kg　　**n** 1488 cm or 14.88 m

**4** 51 kg

## Mastery

**1** **a** 70.2　　**b** 57.5　　**c** 75.9
　　**d** 187.5　　**e** 226.8　　**f** 164.5
　　**g** 109.2　　**h** 236.8　　**i** 136.9

**2** She has $125.65 × 52 = $6533.80 by the end of the year. This means she is $165.20 short of the price of the car.

# UNIT 2: Topic 7

## Practice

**1** **a** 0.04, 4%. Student shades 4 squares.
　　**b** $\frac{23}{100}$, 23%. Student shades 23 squares.
　　**c** $\frac{45}{100}$, 0.45. Student shades 45 squares.
　　**d** 0.4, 40%. Student shades 40 squares.

**2** Other equivalent fractions are acceptable.

| | Fraction | Decimal | Percentage |
|---|---|---|---|
| **a** | $\frac{85}{100}$ | 0.85 | 85% |
| **b** | $\frac{3}{100}$ | 0.03 | 3% |
| **c** | $\frac{7}{100}$ | 0.07 | 7% |
| **d** | $\frac{7}{10}$ | 0.7 | 70% |
| **e** | $\frac{3}{10}$ | 0.3 | 30% |
| **f** | $\frac{65}{100}$ | 0.65 | 65% |
| **g** | $\frac{3}{4}$ | 0.75 | 75% |
| **h** | $\frac{99}{100}$ | 0.99 | 99% |
| **i** | $\frac{9}{10}$ | 0.9 | 90% |
| **j** | $\frac{1}{4}$ | 0.25 | 25% |
| **k** | $\frac{3}{5}$ | 0.6 | 60% |
| **l** | $\frac{18}{100}$ | 0.18 | 18% |

**3** **a** $\frac{1}{2}$ = 50%: True　　**b** 0.07 < 70%: True
　　**c** 0.23 = $\frac{23}{100}$: True　　**d** 75% > $\frac{3}{4}$: False

**e** $\frac{4}{100}$ < 40%: True　**f** 0.89 = 89%: True　**g** $\frac{1}{5}$ > 20%: False　**h** 35% = 0.35: True
**i** 1% = 0.01: True

**4** **a** 65%, 0.7, $\frac{3}{4}$　　**b** 29%, 0.292, $\frac{3}{10}$

## Challenge

**1** Matches are:
- 60%, $\frac{3}{5}$, 0.6　　• 25%, $\frac{1}{4}$, 0.25　　• 20%, $\frac{1}{5}$, 0.2　　• 80%, 0.8, $\frac{4}{5}$　　• 7%, $\frac{7}{100}$, 0.07

The one with no match is 0.7.

**2** Teachers to decide the level of tolerance for the responses. Intended positions are:
Percentages:
25%, 40%, 57%, 72%, 99%
Fractions (Allow equivalents):
$\frac{5}{100}$, $\frac{35}{100}$, $\frac{53}{100}$, $\frac{3}{4}$, $\frac{99}{100}$

**3** Teachers to decide the level of tolerance for the responses. Positions are:
　　**a** square: 9%　　**b** star: 40%　　**c** circle: 58%
　　**d** hexagon: 66%　　**e** pentagon: 91%

**4** Look for students who apply knowledge learned in Unit 2: Topic 3 (Decimal fractions) in order to convert fractions to decimals.
　　**a** square: $\frac{1}{12}$ = 0.08 = 8%　　**b** star: $\frac{5}{12}$ = 0.42 = 42%　　**c** circle: $\frac{7}{12}$ = 0.58 = 58%
　　**d** hexagon: $\frac{8}{12}$ = 0.67 = 67%　　**e** pentagon: $\frac{11}{12}$ = 0.92 = 92%

## Mastery

**1**

| | Land area (nearest $\frac{1}{2}$ million km²) | Area which is forest (nearest $\frac{1}{2}$ million km²) | Area which is forest (nearest 1%) | Area of inland water (nearest 5000 km²) | Area which is inland water (nearest 0.1%) |
|---|---|---|---|---|---|
| Russia | 17 million km² | 8 million km² | 47% | 85 000 km² | 0.5% |
| Brazil | 8.5 million km² | 5 million km² | 59% | 59 000 km² | 0.7% |
| Canada | 10 million km² | 3 million km² | 30% | 900 000 km² | 9% |
| USA | 10 million km² | 3 million km² | 30% | 660 000 km² | 6.6% |
| China | 9.5 million km² | 2 million km² | 21% | 270 000 km² | 2.8% |
| Australia | 7.5 million km² | 1.5 million km² | 20% | 70 000 km² | 0.9% |

**2** Teachers may wish to discuss why figures such as those in the table are rounded. Actual figures may vary according to the research undertaken. For Australia the figures are:
Forest 1 636 780 km² = 21.28%
Inland water: 68 920 km² = 0.89%

# UNIT 3: Topic 1

## Practice

**1** **a** $45 (accept $44.99)
　　**b** $15 (accept $14.99)
　　**c** $14.40 (accept $14.39)
　　**d** $7.50 (accept $7.49)
　　**e** $36 (accept $35.99)
　　**f** $3.60 (accept $3.59)
　　**g** $63 (accept $62.99)
　　**h** $3.75 (accept $3.74)
　　**i** $13.50 (accept $13.49)
　　**j** $12.50 (accept $12.49)

## Challenge

**1** **a** $176.25 (accept $176.24)
　　**b** $105.70
　　**c** $88.15
　　**d** $64 (accept $63.99)
　　**e** $184 (accept $183.99)
　　**f** $788 (accept $787.99)
　　**g** $17 (accept $16.99)
　　**h** $96 (accept $95.99)
　　**i** $153 (accept $152.99)
　　**j** $522.50 (accept $522.49)

## Mastery

**1** The sequences that will give the correct answer are:
Sequence c: 10, %, ×, 135, =
Sequence e: 135, ×, 10, %, =

**2** Estimations may vary.
　　**a** Estimate: $1.60.
　　　　Exact answer: $1.52.
　　**b** Estimate: $1.50.
　　　　Exact answer: $1.47.
　　**c** Estimate: $7.50.
　　　　Exact answer: $7.42(3).
　　**d** Estimate: $16.
　　　　Exact answer: $15.60.
　　**e** Estimate: $1.
　　　　Exact answer: $1.04.
　　**f** Estimate: $9.
　　　　Exact answer: $9.21.
　　**g** Estimate: $750.
　　　　Exact answer: $727.50.
　　**h** Estimate: $600.
　　　　Exact answer: $614.90

**3** $825.55

Oxford University Press

## UNIT 4: Topic 1

### Practice

1  Teacher to check, e.g. "Five sticks are needed for every pentagon".

2  **a**

| Number of pentagons | 1 | 2 | 3 | 4 | 5 | 6 | 7 | 8 | 9 | 10 |
|---|---|---|---|---|---|---|---|---|---|---|
| Number of sticks | 5 | 10 | 15 | 20 | 25 | 30 | 35 | 40 | 45 | 50 |

  **b**  100

3  Student circles: "Start with one stick and then use four sticks for every pentagon".

4  **a**  $5 + 4 \times 2 = 13$
   **b**  $5 + 4 \times 4 = 21$

5  Answers may vary but are likely to be similar to the following:
   "You need six sticks for the first hexagon and five for every other hexagon," or "You need one stick to start off and then five sticks for every hexagon".

### Challenge

1  **a**  Yes: $972 \div 9 = 108$
   **b**  No: $799 \div 9 = 88 \text{ r}7$
   **c**  Yes: $441 \div 9 = 49$
   **d**  Yes: $4284 \div 9 = 476$
   **e**  No: $3849 \div 9 = 427 \text{ r}6$
   **f**  Yes: $123\,723 \div 9 = 13\,747$

2  Student writes any six different digits, the sum of which is 9, then writes the corresponding division problem. For example, $791\,235 \div 9 = 87\,915$.

3  Teacher to check. Correct responses are likely to be similar to the following:

   **a**

---

**b**
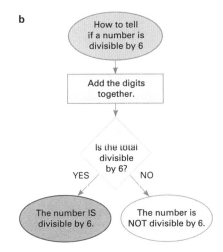

4  **a**  Square it, then add 2
   **b**  4.25 ($1.5 \times 1.5 = 2.25$. Add 2 = 4.25)

### Mastery

1  Look for students who recognise that any number that is divisible by 2 is also divisible by 4 and 8, and do not use them as part of the flow chart. Similarly, numbers that are divisible by 3 are also divisible by 6 and 9. The flow chart is likely to look similar to the following:

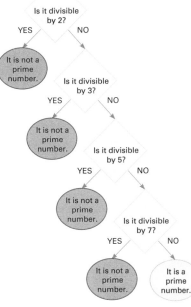

2  2-digit prime numbers are:
   11, 13, 17, 19, 23, 29, 31, 37, 41, 43, 47, 53, 59, 61, 67, 71, 73, 79, 83, 89 and 97

---

## UNIT 4: Topic 2

### Practice

1  **a**  14        **b**  19        **c**  3
   **d**  10        **e**  8         **f**  9
   **g**  21        **h**  14        **i**  $7\frac{1}{2}$

---

   **j**  6         **k**  27        **l**  81
   **m**  81        **n**  29        **o**  1
   **p**  $2\frac{1}{2}$

2
| | |
|---|---|
| **a** $42 \times 3 = (40 \times 3) + (2 \times 3) = 120 + 6 = 126$ |
| **b** $14 \times 6 = (10 \times 6) + (4 \times 6) = 60 + 24 = 84$ |
| **c** $53 \times 5 = (50 \times 5) + (3 \times 5) = 250 + 15 = 265$ |
| **d** $23 \times 9 = (20 \times 9) + (3 \times 9) = 180 + 27 = 207$ |

3  Examples of how the problems might be solved:

| | |
|---|---|
| **a** $25 \times 11 \times 4 = 25 \times 4 \times 11 = 100 \times 11 = 1100$ |
| **b** $50 \times 12 \times 4 = 50 \times 4 \times 12 = 200 \times 12 = 2400$ |
| **c** $8 \times 15 \times 4 = 15 \times 4 \times 8 = 60 \times 8 = 480$ |
| **d** $5 \times 19 \times 2 = 5 \times 2 \times 19 = 10 \times 19 = 190$ |

### Challenge

1
| | Problem | Use opposites | Write the value of ◆ | Check by writing the equation |
|---|---|---|---|---|
| **a** | $◆ \times 7 = 63$ | $◆ = 63 \div 7$ | 9 | $9 \times 7 = 63$ |
| **b** | $◆ - 2\frac{1}{2} = 6$ | $◆ = 6 + 2\frac{1}{2}$ | $8\frac{1}{2}$ | $8\frac{1}{2} - 2\frac{1}{2} = 6$ |
| **c** | $\frac{1}{4}$ of ◆ = 15 | $◆ = 15 \times 4$ | 60 | $\frac{1}{4}$ of 60 = 15 |
| **d** | $◆ \times 100 = 225$ | $◆ = 225 \div 100$ | 2.25 | $2.25 \times 100 = 225$ |
| **e** | $◆ \div 100 = 1.6$ | $◆ = 1.6 \times 100$ | 160 | $160 \div 100 = 1.6$ |
| **f** | $◆ \div 8 = \frac{1}{8}$ | $◆ = \frac{1}{8} \times 8$ | 1 | $1 \div 8 = \frac{1}{8}$ |
| **g** | $◆ \times 1000 = 7225$ | $◆ = 7225 \div 1000$ | 7.225 | $7.225 \times 1000 = 7225$ |

2
| | Problem | Possible substitutes for ◆ | | | | Check by writing the equation |
|---|---|---|---|---|---|---|
| **a** | $◆ \times (2 + 3) = 50$ | 8 | 9 | **10** | 11 | $10 \times (2 + 3) = 50$ $= 10 \times 5 = 50$ |
| **b** | $72 \div ◆ - 5 = 3$ | **9** | 10 | 11 | 12 | $72 \div 9 - 5 = 3$ $= 8 - 5 = 3$ |
| **c** | $1\frac{1}{2} \times ◆ + 6 = 10\frac{1}{2}$ | 2 | **3** | 4 | 5 | $1\frac{1}{2} \times 3 + 6 = 10\frac{1}{2}$ $= 4\frac{1}{2} + 6 = 10\frac{1}{2}$ |
| **d** | $25 \div ◆ + 2.5 = 5$ | 5 | **10** | 15 | 20 | $25 \div 10 + 2.5 = 5$ $= 2.5 + 2.5 = 5$ |
| **e** | $25 \times (10 - ◆) = 150$ | 3 | **4** | 5 | 6 | $25 \times (10 - 4) = 150$ $= 25 \times 6 = 150$ |
| **f** | $◆ \times 2 = 3 + 5^2$ | 12 | 13 | **14** | 15 | $14 \times 2 = 3 + 25$ $= 14 \times 2 = 28$ |
| **g** | $(5 + ◆) \times 10 = 25 \times 3$ | 1.5 | 2 | **2.5** | 3 | $(5 + 2.5) \times 10$ $= 25 \times 3$ $= 7.5 \times 10 = 75$ |

3  **a**  $25 \times 4 = 10^2$        **b**  $1\frac{1}{2} \times 2 = 3$
   **c**  $4 + 2 \times 3 = 10$        **d**  $\frac{1}{2}$ of $6 \times 5 = 15$

### Mastery

1  The correct answer can be found using $(27 + 23) \times 3 = 50 \times 3 = 150$. The other equation, $27 + 23 \times 3$, would give an answer of 96, which is incorrect in the context of this problem.

2  Although the students can easily see that the answer is 151, this is a task to see whether they can identify an equation that will solve the problem. Students who are successful in this are likely to be able to apply this skill to other situations.
   $1 + (27 + 23) \times 3 = 1 + 50 \times 3 = 1 + 150 = 151$

3  In these questions it is also possible to arrive at the answer without writing equations, but teachers will probably wish to encourage their students to do so in order to improve their problem-solving skills.

# ANSWERS

**a** Some students may choose to bracket the numbers being multiplied in order to clarify the equation. Teachers could reassure them that, although this is not necessary, as doing so does not affect the outcome.
$(4 + 2) \times 2 + 3 \times 5 + 2 +$
$2 \times 5 + 2 \times 10$
$= 6 \times 2 + 3 \times 5 + 2 + 2 \times 5 + 2 \times 10$
$= 12 + 15 + 2 + 10 + 20 = 59$
[The first line could be written as $(4 + 2) \times 2 + (3 \times 5) + 2 + (2 \times 5) + (2 \times 10)$ for clarification purposes.]

**b** The operations can be carried out as follows:
$2 \times 5 + \frac{1}{4}$ of $96 + 2 \times 10$
$= 10 + 24 + 20 = 54$.
This leaves $96 - 54 = 42$ pages left to read.

## UNIT 5: Topic 1

### Practice

**1 a**

| cm | mm |
|---|---|
| 2 cm | 20 mm |
| 3.5 cm | 35 mm |
| 4.7 cm | 47 mm |
| 12.3 cm | 123 mm |
| 15 cm | 150 mm |

**b**

| m | cm |
|---|---|
| 4 m | 400 cm |
| 2.5 m | 250 cm |
| 2.32 m | 232 cm |
| 4.35 m | 435 cm |
| 7.5 m | 750 cm |

**c**

| km | m |
|---|---|
| 3 km | 3000 m |
| 14 km | 14 000 m |
| 4.75 km | 4750 m |
| 3.9 km | 3900 m |
| 1.8 km | 1800 m |

**2** Students could be asked to justify other responses.

| | Object | 1ˢᵗ unit | 2ⁿᵈ unit |
|---|---|---|---|
| a | The length of a paper clip | 2.7 cm | 27 mm |
| b | The height of a young student | 127 cm | 1.27 m |
| c | The length of a finger | 7.5 cm | 75 mm |
| d | The height of a ceiling | 2.7 m | 2700 mm |
| e | The length of a street | 2.7 km | 2700 m |

**3** Teachers to decide on amount of tolerance for lengths. Suggest +/–1 mm. It is more important to be consistent across all three columns.

| | mm | cm and mm | cm |
|---|---|---|---|
| a | 35 mm | 3 cm and 5 mm | 3.5 cm |
| b | 64 mm | 6 cm and 4 mm | 6.4 cm |
| c | 58 mm | 5 cm and 8 mm | 5.8 cm |
| d | 86 mm | 8 cm and 6 mm | 8.6 cm |

### Challenge

**1** Student should circle 3 m 50 cm and 350 cm.

**2 a** Estimates may vary, but should be within 1 cm less than 8.5 cm for Line A and 1 cm more than 8.5 cm for Line B.

**b** Teachers to decide on amount of tolerance for lengths. Suggest +/– 1 mm.
Line A: 8.1 cm
Line C: 9.2 cm

---

**3 a** one side    **b** 3 cm

**4** Teachers to decide on amount of tolerance for perimeters. Suggest +/– 1 mm per side.

**a** $2 \times (3.5 \text{ cm} + 1.9 \text{ cm}) = 10.8 \text{ cm}$. Student answers, "I measured two sides".

**b** $6 \times 1.5 \text{ cm} = 9 \text{ cm}$. Student answers, "I measured one side".

**5** Teachers to decide on amount of tolerance for the lengths of the sides and the accuracy of the shape. The aim is for the students to draw a rectangle that is 14.8 cm × 2.6 cm.

### Mastery

**1 a and b** Teacher to check. Students could be asked to show their working out.

**c** Most likely response is 99 cm, but others are possible.

**d** Answers may vary, e.g. the length of the whiteboard.

**2** Students may use other units of length.

**a** Length: 134 cm – 7.53 cm = 126.47 cm
Width: 99 cm – 33.59 cm = 65.41 cm
Height: 120 cm – 56.5 cm = 63.5 cm

**b** Practical activity

**c** Answers may vary, e.g. an armchair.

**3** 2.58 cm or 25.8 mm

**4** A micrometre is a unit of length based on the metre. It is one-millionth of a metre. This means that it is one ten-thousandth of a centimetre. There are 1000 micrometres in 1 mm. The guitar, therefore, is one-hundredth of a millimetre in length. It has guitar strings which are five-hundredths of a micrometre thick.

## UNIT 5: Topic 2

### Practice

**1 a** 18 cm²    **b** 16 cm²    **c** 21 cm²
**d** 6 cm²    **e** 12 cm²

**2** Balcony: 21 m², bedroom: 18 m², bathroom: 9 m², lounge room: 36 m², kitchen/dining room: 35 m²

### Challenge

**1 a** 14 cm²    **b** 24 cm²    **c** 9 cm²
**d** 23 cm²    **e** 42 cm²

**2** Teacher to check. Look for students who use the knowledge of splitting shapes into rectangles to create the two different shapes.

### Mastery

**1** Some students may choose to draw rectangles around the triangles (see Student Book 6, page 93). Others may prefer to imagine the rectangles. Teachers to decide whether students are ready to use a formula for calculating the area of a triangle, e.g. half the length of the base multiplied by the height of the triangle.

**a** 15 cm²    **b** 6 cm²

**2** See also the note for question 1.

**a** 24 cm²    **b** 10 cm²

---

**3** Students could spend as little or as much time as they (or the teacher) desire on this task. Students could share their ideas with each other. Alternatively, the task could be completed as a cooperative group activity.
A 9 m × 9 m enclosure would provide the largest area (81 m²). Drawing different rectangles by increasing the length by a metre and reducing the width by a metre each time would use all the wire, but the area would become smaller and smaller (10 m × 8 m gives a perimeter of 36 m but the area reduces to 80 m²; 11 m × 7 m gives a perimeter of 36 m but further reduces the area to 77 m², and so on.).

## UNIT 5: Topic 3

### Practice

**1 a** 18 centimetre cubes   **b** 18 cm³

**2 a** 80 cm³    **b** 63 cm³    **c** 84 cm³

**3 a**

| Kilolitres | Litres |
|---|---|
| 5 kL | 5000 L |
| 2.5 kL | 2500 L |
| 4.25 kL | 4250 L |
| 3.75 kL | 3750 L |

**b**

| Litres | Millilitres |
|---|---|
| 4 L | 4000 mL |
| 3.5 L | 3500 mL |
| 2.25 L | 2250 mL |
| 9.75 L | 9750 mL |

**c**

| Volume | Capacity |
|---|---|
| 100 cm³ | 100 mL |
| 500 cm³ | 500 mL |
| 175 cm³ | 175 mL |
| 2000 cm³ | 2 L |

**4** 375 mL, 600 mL, $\frac{3}{4}$ L, 1 L 200 mL, 1.25 L, 1375 mL, $1\frac{1}{2}$ L

### Challenge

**1** Teacher to decide on level of accuracy required for the shading. Answers may vary according to the choice of unit of capacity.

**a** 1200 mL or 1.2 L or 1 L 200 mL
**b** 1250 mL or 1.25 L or 1 L 250 mL
**c** 2000 mL or 2 L
**d** 1875 mL or 1.875 L or 1 L 875 mL
**e** 1975 mL or 1.975 L or 1 L 975 mL
**f** 1750 mL or 1.75 L or 1 L 750 mL

**2** Answers may vary. Successful responses are likely to include the fact that there are 10 cubes on the bottom layer and three layers altogether, making 30 cubes in total.

**3** Likely responses are:
L: 10 cm, W: 3 cm, H: 1 cm; or
L: 15 cm, W: 2 cm, H: 1 cm; or
L: 30 cm, W: 1 cm, H: 1 cm

Oxford University Press

Teachers may need to discuss with students whether a prism with dimensions of:
L: 3 cm, W: 2 cm, H: 5 cm
is in fact different from one with dimensions of:
L: 5 cm, W: 2 cm, H: 3 cm.

## Mastery

1  Some facts about megalitres and gigalitres:
**Megalitre — ML**
- 1 000 000 L (one million litres) = 1 ML
- It is the amount of water that would fit into a 10 m cube, i.e. 1000 m³.
- It would fill almost half of an Olympic–size swimming pool (50 m long × 25 m wide × 2 m deep).
- Farmers are supplied with water in megalitres.
- At 100% full, the Hume Dam can hold over 3 000 000 ML – approximately five times the volume of Sydney Harbour.

**Gigalitre — GL**
- 1 000 000 000 litres (one billion litres) = 1 GL
- It is the amount of water that would fit into a 100 m cube, i.e. 1 000 000 m³.
- Water flowing in rivers is measured in gigalitres.
- The volume of Sydney Harbour is approximately 560 GL.

2  This could form part of a group or class challenge to find as many different rectangular prisms as possible, each with a volume of 24 cm³. Teachers may wish to discuss with students whether a prism in a different orientation (e.g. using a different face for the base) actually constitutes a different rectangular prism.
Responses should include (in various orientations):
L: 4 cm, W: 2 cm, H: 3 cm
L: 6 cm, W: 2 cm, H: 2 cm
L: 12 cm, W: 2 cm, H: 1 cm
L: 24 cm, W: 1 cm, H: 1 cm
L: 8 cm, W: 3 cm, H: 1 cm
Note: The responses do not include dimensions of prisms that have edges measured in fractions of a centimetre, but this could be an interesting discussion point for students.

## UNIT 5: Topic 4

## Practice

1  a

| Tonnes | Kilograms |
| --- | --- |
| 2 t | 2000 kg |
| 3.5 t | 3500 kg |
| 4.25 t | 4250 kg |
| 5.175 t | 5175 kg |
| 0.975 t | 975 kg |

b

| Kilograms | Grams |
| --- | --- |
| 5 kg | 5000 g |
| 3.5 kg | 3500 g |
| 0.75 kg | 750 g |
| 0.45 kg | 450 g |
| 3.07 kg | 3070 g |

c

| Grams | Milligrams |
| --- | --- |
| 3 g | 3000 mg |
| 2.25 g | 2250 mg |
| 1.5 mg | 1500 mg |
| 0.735 g | 735 mg |
| 0.001 g | 1 mg |

2  Other ways of writing the masses are possible (e.g. using fractions).
a  3 kg 750 g or 3.75 kg
b  750 g or 0.75 kg
c  2 kg 300 g or 2.3 kg

3  Students could be asked to justify an alternative response.
a  B    b  A    c  C    d  B or C

4  Teacher to decide on level of accuracy. There are 250 g increments on the scale, so the pointer should be closer to 4 kg 500 g than to 4 kg 250 g.

## Challenge

1  Estimates should be around the ones shown. Teacher to decide on level of accuracy required. However, different ways of writing the mass should correspond with each other, e.g.
a  2.7 kg, 2 kg 700 g, 2700 g, 2 $\frac{7}{10}$ kg
b  870 g    c  1.225 kg

2  a  C, B, A
b  Any mass between the 1300 kg and 1305 kg is correct but units should correspond to each other. Possible answers (without fractions of a kilogram):
1.301 t, 1 t 301 kg, 1301 kg
1.302 t, 1 t 302 kg, 1302 kg
1.303 t, 1 t 303 kg, 1303 kg
1.304 t, 1 t 304 kg, 1304 kg
c  0.265 t or 265 kg

3  Practical activity. It is doubtful that, unless the mass of each object is recorded on it, the students will estimate to exactly 1 kg. The main aim is for students to show their proficiency in estimating mass.

## Mastery

1  a  330 kg + 570 kg + 0.62 t + 1.15 t = 2.67 t = Truck C
b  0.475 t + 1.45 t + 255 kg + 345 kg = 2.525 t = Truck B
c  0.075 t + 0.4 t + 1 t 425 kg + 900 kg = 2.8 t = Truck A

2  a  9.5 kg, 9 kg 500 g, 9500 g
b  9500 g ÷ 12 = 792 g (to nearest gram)
c  Look for students who work through the problem logically and who use appropriate masses. An easy, although not a very likely, real-life solution would be to start at 791 and subtract/add one each time up to the 11th apple.
791 + 790 + 792 + 789 + 793 + 788 + 794 + 787 + 795 + 786 + 796 = 8701 g
9500 g – 8701 g = 799 g for the 12th apple.

## UNIT 5: Topic 5

## Practice

1  a  Clock to show 4:34, 1634    b  2:34 am, 0234    c  Clock to show 1:13, 1:13 pm
d  7:23 am, 0723    e  Clock to show 11:47, 1147    f  Clock to show 12:03, 12:03 am

2

Key: 1 year

G   O   L   D   R   U   S   H

1851  1852  1853  1854  1855  1856  1857  1858  1859  1860  1861  1862  1863  1864

## Challenge

1  a  31 minutes    b  06:24    c  9 minutes    d  07:16
e  06:13 (Because the 06:31 does not stop at Lapstone.)

2  a  1:40 pm    b  4 hours and 41 minutes (07:29 to 12:10)    c  7 minutes
d  The journey from Penrith to Springwood is shorter by one minute.

## Mastery

1  a  The order is C, E, F, A, D, B.
b  Teacher to decide on level of accuracy that is needed for the arrows. The midday markers are shown.

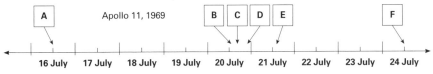

A   Apollo 11, 1969   B  C  D  E   F

16 July  17 July  18 July  19 July  20 July  21 July  22 July  23 July  24 July

c  2256    d  6 hours, 38 minutes    e  4 days, 4 hours and 10 minutes
f  8 days, 3 hours and 18 minutes

## UNIT 6: Topic 1

### Practice

1 Regular shapes to have horizontal stripes. Irregular shapes to have vertical stripes.
   a irregular triangle
   b regular pentagon
   c irregular quadrilateral
   d irregular octagon
   e regular heptagon (septagon)
   f irregular hexagon
2 Teacher to check. Look for students who note that, because the shape does not have straight sides, it cannot be called a polygon.
3 Students circle, "It is an irregular quadrilateral with one pair of parallel sides".
4 Students draw a square.

### Challenge

1 Answers may vary, but look for students who describe the triangle's properties rather than, say, its size.
   a equilateral (All three sides are of equal length. All three angles are the same size.)
   b right-angled (The three sides are different lengths. It has one right angle.)
   c isosceles (Two of the sides and two of the angles are the same size.)
   d scalene (None of the sides and none of the angles are the same size.)
   e isosceles and right-angled (Two of the sides and two of the angles are the same size. It has one right angle)
2 Answers will vary, as any one of the shapes could be "the odd one out". However, look for students who justify their choice(s) by virtue of each shape's properties rather than its appearance. Responses could include, but are not limited to:
   a This is the only shape with two pairs of parallel sides.
   b This is the only shape with two equal sides and two equal angles.
   c This is only shape that is not a polygon.
   d This is the only regular polygon.
   e This is the only polygon with reflex angles.
3 Teachers to decide on level of accuracy. Students should draw a parallelogram with two sides 9 cm long.

### Mastery

1 Teacher to decide on level of accuracy required, and to check that students label the circle correctly.
2 Practical activity.

## UNIT 6: Topic 2

### Practice

1 a cube, prism
   b sphere, other 3D object
   c hexagonal pyramid, pyramid
   d rectangular prism, prism
   e cone, other 3D object
   f hexagonal prism, prism

2

| | Shape of the base(s) | Number of bases |
|---|---|---|
| a | triangular | 1 |
| b | triangular | 2 |
| c | square | 2 |
| d | pentagonal | 2 |

### Challenge

Some students will need extra paper on which to try out their sketches. It is important that students do not feel that they have failed if they cannot master the skill of sketching a 3D object at the first, or even the fifth, attempt.

1 Teachers to decide on the level of accuracy required.
2 Teachers will note that there is no indication of the size of the objects. The main aim of this task is for students to attempt a representation of the object. Teachers to use their professional judgement about the degree of accuracy and proportion that is expected.

### Mastery

1 Students could practise by making and cutting a cross-section using modelling clay. Others will be able to visualise the cross-section in the abstract. Students should draw:
   a a rectangle       b a circle
   c a hexagon        d a pentagon
   e a circle          f a square
2 Students should draw:
   a an oval
   b a rectangle
3 The 3D objects are:
   a a cylinder
   b a cone
   Teacher to check explanation.

## UNIT 7: Topic 1

### Practice

1 a 70°, acute
   b 110°, obtuse
   c 155°, obtuse
2 We know that the angle is 330° degrees, because 360° – 30° = 330°.
3 a 310°            b 325°
   c 345°

### Challenge

1 Suggested tolerance is +/– 1°.
   a 122°
   b 78°
2 Teacher to decide on desired level of accuracy. Suggested tolerance is +/– 2°.
3 a 85°
   b 117°
   c 88°
   d 322°
   e 124°, 56°

### Mastery

1 Teachers may wish to allow a tolerance of +/– 1° but corresponding, supplementary and opposite angles must reflect the size of the initial angle.
   Angle a: 123°
   Angles c, f, h, i, k n, p: 123°
   Angles b, d, e, g, j, l, m, o: 57°
2 Students may wish to carry out research before commencing. The way to work out the sum of the angles in a polygon is to start with the sum of the angles in a triangle, which we already know to be 180°. Each time an extra side is added, we simply add 180°.
   So, quadrilateral = 360°, pentagon = 540°, hexagon = 720° and so on.
   If students draw irregular polygons (and probably even if they draw regular ones), they may need to "tweak" the measurements of the angle sizes in order for the pattern to be maintained.

## UNIT 8: Topic 1

### Practice

1

2 a rotation
   b reflection
   c translation
3 Student circles, "The shape could have been translated, reflected or rotated".

### Challenge

1
2
3
4

### Mastery

1 This could become a "think, pair, share" activity, with students identifying features, such as the parting in the hair, that make the first picture asymmetrical.

Oxford University Press

**2** Teachers to decide on level of accuracy that is required for this difficult activity. The image could be duplicated for students who would like more than one attempt at this task.

**3** Practical activity

## UNIT 8: Topic 2

### Practice

**1 a** (0,7)  **b** (−8,1)  **c** (−5,−8)
 **d** (5,−8)  **e** (8,1)

**2** The finishing point must be the same as the starting point in order to close the shape.
(−8, 9) → (−5,9) → (−4, 7) → (−5,5) → (−8,5) → (−9,7) → (−8,9)

**3** Students join the dots to make a pentagon. Answers for drawing the pentagon could vary depending on the starting point. However, the finishing point must be the same as the starting point in order to close the shape.
(0,7) → (−8,1) → (−5,−8) → (5,−8) → (8,1) → (0,7)

**4** Students draw an isosceles triangle of their choice. Answers for drawing the triangle will vary depending on the points chosen. Teachers could ask students to follow each other's instructions to check whether the points have been plotted accurately.

### Challenge

**1 a** (−9, 6) → (−7,7) → (−5,7) → (−4,8) → (−4,6) → (−9,6)
 **b** See illustration below for answer.
 **c** (9, 6) → (7,7) → (5,7) → (4,8) → (4,6) → (9,6)

**2** See illustration below for answer.

**3 a**

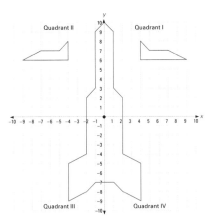

 **b** (0,10) → (1,9) → (1,3) → (2,2) → (2,−4) → (4,−5) → (4,−9) → (2,−8) → (1,−7) → (0,−7)

### Mastery

**1**
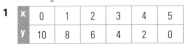

| x | 0 | 1 | 2 | 3 | 4 | 5 |
|---|---|---|---|---|---|---|
| y | 10 | 8 | 6 | 4 | 2 | 0 |

**2 a**

| x | 0 | 1 | 2 | 3 | 4 | 5 | 6 | 7 | 8 | 9 | 10 |
|---|---|---|---|---|---|---|---|---|---|---|---|
| y | 10 | 8 | 6 | 4 | 2 | 0 | −2 | −4 | −6 | −8 | −10 |

 **b** Quadrant IV
 **c** See illustration to the right.

**3 a**

| x | 0 | 1 | 2 | 3 | 4 | 5 | 6 | 7 | 8 | 9 | 10 |
|---|---|---|---|---|---|---|---|---|---|---|---|
| y | −10 | −8 | −6 | −4 | −2 | 0 | 2 | 4 | 6 | 8 | 10 |

 **b** See illustration to the right.

**c**

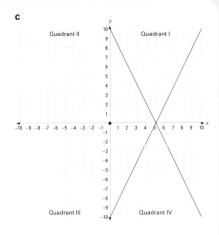

## UNIT 9: Topic 1

### Practice

**1 a**

| Frequency table: The number of people who came to visit the school garden | | | | | |
|---|---|---|---|---|---|
| Day | Monday | Tuesday | Wednesday | Thursday | Friday |
| Number | 11 | 14 | 21 | 22 | 29 |

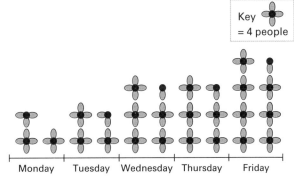

Key ✿ = 4 people

 **b** 97

**2 a–b**

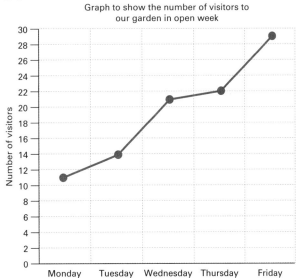

 **c** Answers may vary, but should reflect the fact that the number of visitors grew day by day.
 **d** This could form part of a class or group discussion. Some might suggest that the number would be greater than 29 because of the trend, whereas others might suggest that the sixth day would be a Saturday and that there would perhaps be fewer visitors.

# ANSWERS

## Challenge

1 a China
  b Answers will vary. Look for students who use their acquired skills to interpret the data, e.g. "The population of the USA is greater than the combined populations of Russia, France and Australia".

2 Answers may vary but are likely to reflect the fact that the approximate size of the populations can be identified on the column graph.
  b 1 400 000 000 (1.4 billion)
  c Estimates need to be around 300 000 000. (Data used for the graph was 323 million.)

3 Look for students who use their acquired skills to interpret the data. Answers will vary, e.g. "The pet cat population of the USA is almost half the total pet cat population of all five countries".

4 The statement is reasonably accurate because 75 million is a quarter of 300 million. Actual data (rounded at the time of writing) used for the graphs was:

| Country | Cats | People |
|---|---|---|
| USA | 76 million | 323 million |
| China | 53 million | 1400 million |
| Russia | 13 million | 144 million |
| France | 10 million | 67 million |
| Australia | 3 million | 25 million |

## Mastery

1 When students are constructing their own graphs, especially if large numbers are concerned, it is important that they be comfortable with rounding to an appropriate number and using a suitable scale for the axes. Teachers may wish to encourage students who undertake any of the activities on this page to discuss the outline of their graphs before they start.

Depending on the experience and ability levels of the students, using digital media to produce graphs can be highly beneficial.
  a The gold production for Russia was rounded to the nearest thousand.
  b Answers may vary, but are likely to be along the lines of "to save space and/or time". Note that the key to the chart indicated that the number had been abbreviated.

If students are entering large numbers into graphing programs, they will find it easier to abbreviate the numbers (by the same amount) for the data, e.g.

| China | 362 |
|---|---|
| Australia | 258 |
| USA | 234 |
| Russia | 199 |
| South Africa | 181 |

instead of:

| China | 362 000 |
|---|---|
| Australia | 258 000 |
| USA | 234 000 |
| Russia | 199 000 |
| South Africa | 181 000 |

would result in the same proportions on the graph itself.

2 Practical activity

## UNIT 9: Topic 2

### Practice

1 a Answer should be around 75%.
  b 2

2 primary

3 Answers may vary. Look for appropriate responses, e.g. "Do you think that we should be allowed to use our mobile phones in class?"

4 a "Should Year 6 students be allowed to use smartphones in classrooms?"
  b Teacher to check reasoning. Students could share their views with each other.

5 Answers may vary but the total of the two should be 100%. Intended response is 45% = No and 55% = Yes, but answers around these figures are acceptable.

### Challenge

1 Teacher to check. Students could discuss this question in groups. Likely answer is that the question is misleading.

2 Teacher to check. Students could discuss this question in groups. Intended answer is that it is "based on fact and could possibly be true".

3 Sample survey

4 Answers may vary. Students could discuss this question in groups. Answers may range from, "It is not fair because it doesn't say how many students were surveyed", to "It is fair because it sounds as though it could be true".

5 a Answers may vary, but should reflect an overwhelming No response. Intended response is 3:99 (Yes: No).
  b The survey question uses the word "play" whereas the graph title has the word "use". This was not a fair representation of the data.

### Mastery

1–2 This page could form the basis of small- or large-group discussion. Answers will vary, and it could be useful for students to write down their ideas and opinions and share them with each other.

3 a The actual percentage was 18%.
  b There has been a marked increase in smartphone ownership. Students could be encouraged to share their ideas and opinions with each other.
  c A reasonable estimate would be over 90%, but it is doubtful that a figure of 100% would ever be reached.
  d Teacher to decide whether this should take place. This could form part of a cooperative group activity.

## UNIT 10: Topic 1

### Practice

1 a 1 in 8    b 25%    c 0.75
  d 12.5%    e red

2 Answers will vary. Teachers might wish to ask students to justify their responses.

3 Students circle $\frac{1}{5}$, 20%, 0.2, 1 in 5, $\frac{2}{10}$.

4 Answers may vary but should be appropriate and are likely to be between 75% and 90%. Students could be asked to justify other responses.

### Challenge

1 a Students colour the following numbers of sectors:
  • green: 4  • red: 6  • purple: 1
  • yellow: 4  • blue: 3
  b There should be 2 white sectors out of 20. These should be described as $\frac{1}{10}$ (or $\frac{2}{20}$), 0.1, 10%.

2 a $\frac{2}{3}$    b 75%    c $\frac{1}{5}$
  d $\frac{7}{12}$    e 0.25    f 0.3̇3̇

3 Students could justify other answers but, based on probability values, the following are likely outcomes. Look for students who apply their knowledge of equivalent fractions.

| Bag | Fractions red: green | My prediction after 50 have been taken out | |
|---|---|---|---|
| A | $\frac{4}{5} : \frac{1}{5}$ | Red: 40 | Green: 10 |
| B | $\frac{1}{10} : \frac{9}{10}$ | Red: 5 | Green: 45 |
| C | $\frac{3}{10} : \frac{7}{10}$ | Red: 15 | Green: 35 |
| D | $\frac{3}{5} : \frac{2}{5}$ | Red: 30 | Green: 20 |

### Mastery

1 Teachers (and students) to decide whether fractions, decimals or percentages should be used for the values. In most cases fractions will be most appropriate because of the total of 52. Equivalent decimals, fractions or percentages are acceptable.
  a 50%
  b $\frac{1}{13}$
  c $\frac{20}{52} = \frac{10}{26} = \frac{5}{13}$
  d $\frac{1}{26}$
  e $\frac{1}{4}$ or 25%
  f $\frac{12}{52} = \frac{6}{26} = \frac{3}{13}$
  g $\frac{8}{52} = \frac{4}{26} = \frac{2}{13}$
  h $\frac{16}{52} = \frac{8}{26} = \frac{4}{13}$

2 a 1 in 13
  b As there are 16 cards worth 10 and there are 51 cards left, the chance is 16 out of 51.
  c $\frac{16}{51} = 31\%$ (to the nearest whole number)

3 Depending on the interest level of the students (and teacher) this activity could be as short or as long as desired. Students could, for example, research the probability value of choosing two particular cards by searching online.

Oxford University Press

## UNIT 10: Topic 2

### Practice

**1** **a** 25%  **b** 0.5  **c** $\frac{3}{4}$

**2** **a** 5  **b** $\frac{1}{4}$

    **c** Practical activity. Teachers could encourage students to record the fractions in their lowest equivalent forms.

    **d** Teacher to check. Students are likely to comment that the difference between the prediction and reality is because of the "element of chance" in the game.

**3** **a** Predictions are likely to be half-and-half for each type.

    **b** Practical activity. Teachers could encourage students to record the fractions in their lowest equivalent forms.

    **c** As there are only two possible outcomes, their predictions are likely to be closer than in question 2.

### Challenge

**1** Students may choose to express the probability value as a fraction, a decimal or a percentage. Equivalents of the following are acceptable:

    **a** $\frac{4}{5}$

    **b** $\frac{1}{5}$

    **c** $\frac{2}{5}$

**2** **a** • a queen: 16  • a black queen: 8
       • Jack of Diamonds: 4

    **b** Practical activity. Teachers could encourage students to record the fractions in their lowest equivalent forms.

    **c** Because of the chance factor, results are likely to vary from student to student.

**3** Because there are 10 cards, the percentages should be easy for students to calculate:

    **a** 40%

    **b** 30%

    **c** 20%

    **d** 10%

    **e** 50%

    **f** 20%

**4** **a** lower $(\frac{4}{11})$

    **b** lower $(\frac{3}{11})$

    **c** lower $(\frac{2}{11})$

    **d** higher $(\frac{2}{11})$

    **e** higher $(\frac{6}{11})$

    **f** lower $(\frac{1}{11})$

**5** **a** $\frac{4}{11} = 0.36 = 36\%$

    **b** $\frac{3}{11} = 0.27 = 27\%$

    **c** $\frac{2}{11} = 0.18 = 18\%$

    **d** $\frac{2}{11} = 0.18 = 18\%$

    **e** $\frac{6}{11} = 0.55$ (accept 0.54) $= 55\%$ (54%)

    **f** $\frac{1}{11} = 0.09 = 9\%$

### Mastery

**1** Practical activity. Teachers may wish students to discuss the rules for the game after playing it a few times in order to make it more difficult, more simple or more interesting.

**2** This would make a good cooperative group activity.

Oxford University Press